MW00713518

Global Sustainability as a Business Imperative

THE PALGRAVE SERIES ON GLOBAL SUSTAINABILITY

Series Editors:

James A. F. Stoner, *Professor of Management Systems and Chair Holder, James A. F. Stoner Chair in Global Sustainability at the Graduate School of Business, Fordham University, New York*

Charles Wankel, *Associate Professor of Management at St. John's University, New York*

The global environmental crisis has been called "the greatest challenge ever faced by our species." In turn, the required changes in business' raison d'être and impacts on the world will be the most profound change in the concept and practice of business ever accomplished. The transformation of business practice must capture and increase for-profit businesses' positive contributions while decreasing and eventually eliminating the harmful effects businesses have on the physical, economic, cultural, social, and political environments of the world. The volumes in *The Palgrave Series on Global Sustainability* will address the transformations that are occurring and need to occur on a global basis in population, consumption, and production required to achieve a sustainable world—with particular attention to the ways business institutions and business leaders can continue to make their contributions to the world while moving from being "a part of the problem to being a part of the solution." The actions businesses can take start with the recognition of business' impacts on the physical, economic, cultural, social, and political environments of the world and include (1) developing business models grounded in sustainability, (2) energizing and motivating businesses and their stakeholders, (3) proactively transforming the environment of business by business, and (4) contributing to reconceptualizing the current dominant paradigms of economic, social, political, and spiritual life. The series will consist of volumes designed to offer the latest cutting-edge research and knowledge about how to forge the future of business organizations in a sustainable world.

Global Sustainability as a Business Imperative
Edited by James A. F. Stoner and Charles Wankel

Global Sustainability as a Business Imperative

Edited by James A. F. Stoner and Charles Wankel

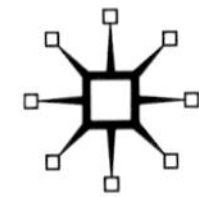

GLOBAL SUSTAINABILITY AS A BUSINESS IMPERATIVE

First published in 2010 by
PALGRAVE MACMILLAN® in the United States – a division of
St. Martin's Press LLC,
175 Fifth Avenue, New York, NY 10010.

Where this book is distributed in the UK, Europe and the rest of the world, this is by Palgrave Macmillan, a division of Macmillan Publishers Limited, registered in England, company number 785998, of Houndmills, Basingstoke, Hampshire RG21 6XS.

PALGRAVE MACMILLAN is the global academic imprint of the above companies and has companies and representatives throughout the world.

ISBN: 978–0–230–10281–1

Library of Congress Cataloging-in-Publication Data

Global sustainability as a business imperative / edited by James A.F. Stoner and Charles Wankel.
p. cm. — (Palgrave Series on global sustainability)
Includes bibliographical references and index.
ISBN 978–0–230–10281–1
1. Sustainable development. 2. Business enterprises—Environmental aspects. 3. Social responsibility of business. I. Stoner, James Arthur Finch, 1935– II. Wankel, Charles.
HC79.E5.G597148 2010
658.4'083—dc22
2010019467

A catalogue record of the book is available from the British Library.

Design by MPS Limited, A Macmillan Company

First edition: December 2010

D 10 9 8 7 6 5 4 3 2

Printed in the United States of America

Contents

Notes on the Contributors

Javier Aguilera-Caracuel is a Graduate at the University of Granada with a degree in General Law and Business Administration in 2007, where he is completing his doctorate in economics and management with a dissertation on Internationalization of Firms and Environmental Management. He is a member of the research group Innovation, Sustainability and Development (ISDE). His research interests include the internationalization of firms, environmental management, and corporate social responsibility.

Cynthia Aguirre received her MBA with a double emphasis in Marketing Management and Entrepreneurship in 2009. Her current position is in the Division of Academic Affairs at Loyola Marymount University. A previous position was in Univision Communications, Inc. at their flagship station in Los Angeles.

J. Alberto Aragón-Correa, Professor of Strategic Management at the University of Granada and Visiting Professor at the University of California, Berkeley, received his PhD in Business Administration and Economics from the University of Seville. He has published in top-tier journals including the *Academy of Management Journal, Academy of Management Review, British Journal of Management, Long Range Planning, Journal of Business Research, Ecological Economics, Sustainable Development*, among others. He has served at the University of Granada as Deputy Dean in the School of Economics and Business and Chair of the Department of Management. He has headed a number of research projects supported by the European Commission on integrating sustainability into universities.

Eugênio Ávila Pedrozo, Professor in the Post Graduate Program Center for Research of Studies in Agribusiness (PPGAgronegócios/CEPAN) and Management (PPGA/EA) at Federal University of Rio Grande do Sul (UFRGS), received his PhD in Management from the Institut National Polytechnique de Lorraine, Nancy, France. He has served as Coordinator of the Organizations

Group Study (GESTOR) and was a Director of CEPAN. He is currently a Department Chief of the DCA (Administration Science Department). His research areas include: strategy, interorganizational relations, agribusiness, complexity, sustainable development, interdisciplinary, and innovation.

Denise Barros de Azevedo received his PhD at the Post Graduate Program Center for Research and Studies in Agribusiness (PPGA/CEPAN) of the Federal University of Rio Grande do Sul (UFRGS) Porto Alegre, RS, Brazil. He received his MSc in Rural Economy in 1998 from the Federal University of Viçosa. Denise participated in GESTOR. His research areas include Stakeholder Theory, interorganizational relations, social network analysis, environmental management, agribusiness, climatic change agribusiness, convention theory, and agribusiness environmental business dialogues. He was President and founder of the Itumbiara Environmental Committee, Goiás, Brazil, 1999 to 2002.

Alessio Cavicchi received his PhD in Economics of Food and Environmental Resources from the University of Naples "Parthenope." He is a Researcher in Agricultural Economics at the University of Macerata (Italy). His main fields of interest and research are consumer food choice, economics of food quality and safety, sustainable tourism and innovation in the agrofood sector. Currently he teaches Rural Development Policy and Economics and Marketing of Food Quality in the degree of Management of Sustainable Tourism Systems. His works have been published in several journals, including *Food Quality and Preference, British Food Journal, Journal of Agricultural Economics, Agribusiness: an International Journal, International Journal of Wine Business Research and Food Economics.*

Avrath Chadha completed his PhD at the Swiss Federal Institute of Technology (ETH), Zurich, Switzerland. His research interests include environmental technologies, particularly polymers made from renewable resources and the management of radical innovation processes. His research work has been presented at the European Group for Organizational Studies (EGOS) as well as at the Academy of Management conference. He currently works with the CEO of Hoerbiger Holding, a diversified multinational technology firm based in Zug, Switzerland.

Guilherme Cunha Malafaia received his PhD in Agribusiness from the Federal University of Rio Grande do Sul (UFRGS). He is Professor and Researcher in the Post Graduate Program in Administration at the University of Caxias do Sul, Brazil. His research areas include agribusiness, sustainable value chain management, local agrifood systems, collective action, coopetition, resource-based view, and convention theory.

Javier Delgado-Ceballos earned his PhD from the University of Granada, where he teaches courses in International Management in the Management Department. His research focuses on connections between organizations and environmental management, stakeholder management, business strategy, and government policy related to sustainability. He has been a visiting scholar at the University St. Gallen and the University of Cyprus.

Blanca Luisa Delgado-Márquez is currently a Lecturer in the Department of International and Spanish Economics in the Faculty of Economics and Business of the University of Granada, from which she received her PhD in Economics and Business. Her research interests include the integration of sustainability into management education.

Nuria E. Hurtado-Torres is an Associate Professor at the University of Granada, from which she received her PhD in Business Administration and Economics. She has been the Coordinator for the Degree in Business Administration in the School of Economics and Business and an officer of the Department of Management at the University of Granada. She has published in top journals including the *Journal of Environmental Management* and the *Journal of Organizational Change Management*. She has participated in different research projects supported by the Spanish Ministry of Education analyzing how to integrate sustainability into universities. Her research interests include the connections between the natural environment and international strategy.

Aileen M. Ionescu-Somers holds a PhD in Business Administration from University College Cork, Ireland. She directs the Center for Corporate Sustainability Management (CSM), at IMD business school in Switzerland. Dr. Ionescu-Somers is an expert in corporate sustainability management and has coordinated several large-scale research projects on, for example, the business case for sustainability, stakeholder views of corporate sustainability performance and sustainability partnerships, publishing many resulting articles and books. Recently she has published on business logic for corporate sustainability in the food and beverage industry with Palgrave Macmillan. Prior to IMD, Dr. Ionescu-Somers oversaw the project and program operations at the WWF, the international conservation NGO. She also managed the Africa/Madagascar and Latin America/Caribbean regional programs at WWF.

Silvester Ivanaj is Associate Professor of Information Systems at ICN Business School, Nancy, France. He received his PhD in Applied Electrochemistry from the Institut National Polytechnique de Lorraine (INPL),

France. Before joining the ICN Business School, he was an environmental consultant. His research interests focus on information systems and sustainability assessment methods.

Vera Ivanaj is an Associate Professor of Management Science in the Chemical Engineering School (ENSIC) of the Institut National Polytechnique de Lorraine (INPL), University of Nancy, France. She received her PhD in Management Science from the University of Nancy II. Prior to joining ENSIC, she was a faculty member of the Business Administration Institute at the University of Metz, where she was in charge of the MBA program. She teaches courses in Strategic Decision Making, Change Management, Human Resource Management, Project management, Conflict Resolution, Corporate Culture and Leadership for managers and engineers. She has published several articles and chapters on enhancing capabilities of organizations to take decisions and to perform more effectively. Her current research interests include strategic decision making, sustainable development, management education, and diversity.

Richard H. Jones, Deputy Executive Director of the International Energy Agency, has a PhD in Business/Statistics from the University of Wisconsin. Ambassador Jones served as the American Ambassador to Israel (2005–2008), Kuwait (2001–2004), Kazakhstan (1998–2001), and Lebanon (1996–1998). During these and earlier assignments he gained a world-class sophistication about oil and gas as well as other energy, economic, and environmental issues.

Candace A. Martinez, Assistant Professor of International Business in the Boeing Institute of International Business at Saint Louis University, was awarded her PhD in strategic management from the University of Illinois at Urbana-Champaign. Her research has focused on the influence of formal and informal norms in institutional environments across countries and, more recently, the role of institutions in governing the informal waste collection/recycling sector in Latin America. Dr. Martinez has conducted field research in Cuba, Central America, and Brazil. She is currently teaching international business courses that allow her to introduce various dimensions of the informal waste-picker/scavenger sector to undergraduates.

Tammy E. Newmark has headed the EcoEnterprises Fund for the Nature Conservancy since 1998 and serves as the fund's President and Treasurer. Newmark is presently launching EcoEnterprises Fund II. Before joining The Nature Conservancy, Newmark led Technoserve, Inc.'s environmental business advisory services in Latin America and Africa. Prior to that, she was a founding officer of Environmental Enterprises Assistance Fund, the first nonprofit venture capital fund that specialized in renewable energy, clean technology, and green investments in emerging markets. At this fund, she established investment groups: Yayasan Bina Usaha Lingkungan and Preferred Energy Investments in Indonesia and the Philippines, respectively.

Natalia Ortiz-de-Mandojana is completing her doctorate in the Management Department of the University of Granada. She is a member of the research group Innovation, Sustainability and Development (ISDE). She was awarded the Regional Government of Andalusia's prize for highest accomplishment for her undergraduate degree program at the University of Granada, from which she was also awarded a law degree. Her dissertation is on Corporate Governance and Environmental Management.

Jacob Park, Associate Professor of Business Strategy and Sustainability at Green Mountain College in Vermont, specializes in the teaching and research of global environment and business strategy, corporate social responsibility, and social and environmental entrepreneurship and innovation with a special expertise and interest in Japan, China, and the Asia-Pacific region. A member of the Steering Committee of the U.S. Social Investment Forum's International Working Group, he has worked as an ethical research consultant with Green Cay Asset Management, a socially responsible hedge fund/financial investment company, and as a senior research consultant of a Japanese and Asian equity specialist in the Governance and Social Responsibility Investment Group of ISIS Asset Management (now F&C Asset Management), a London-based investment company.

Benedetto Rocchi, Lecturer at the School of Agriculture of the University of Florence, from which he obtained his PhD in Forest Economics and Planning. He published several papers on Italian and international refereed journals including *European Review of Agricultural Economics, Review of Agricultural and Environmental Studies, International Journal of Sustainable Development, International Journal of Wine Marketing and Food Economics.*

His main fields of research are the macroeconomic models for policy analysis and the economics of food supply chain. He devoted a relevant part of his activity, joining with international and national research projects on food safety, trust in the food supply chain, and the definition of quality in supply chain of traditional foods.

Antonio Rueda-Manzanares, currently teaches Strategic Management at the University of Granada, from which he received his PhD. His research focuses on various topics relating to organizations and the natural environment including the development and deployment of organizational capabilities, and the stakeholder engagement to generate proactive corporate sustainability strategies and competitive advantage.

John L. Stancil, Professor of Accountancy at Florida Southern College, received his DBA from the University of Memphis, MBA from the University of Georgia, and a BS in Accounting from Mars Hill College. He currently teaches courses in Taxation and Cost/Managerial Accounting. His research interests are primarily in the areas of tax policy issues and sustainability.

James A. F. Stoner is Professor of Management Systems at Fordham University's Graduate School of Business and Chairholder of the James A. F. Stoner Chair in Global Sustainability—a chair endowed in Jim's name by one of Jim's students (Brent Martini) and his father (Bob Martini) in acknowledgment of Jim's teaching and research at Fordham and his contributions to their and their company's work. He earned his BS in Engineering Science at Antioch College in Yellow Springs, Ohio, and his MS and PhD in Industrial Management at MIT, in Cambridge, MA. He has published articles in such journals and periodicals as *Academy of Management Review, Harvard Business Review, Journal of Development Studies, Personnel Psychology, and Journal of Experimental Social Psychology*, and has authored, coauthored, and coedited somewhere around 16 to 20 books. Jim's current projects and interests focus on finding ways to move toward a sustainable world. He has taught managers, executives, MBA and undergraduate students in at least a dozen countries including Brazil, Ethiopia, Iran, Ireland, Japan, and Russia. He has won teaching awards at Fordham and has consulted with a broad range of companies, including Bell Labs, Richardson-Merrill, Bergen Brunswick, and Arthur D. Little, Inc. E mail: stoner@fordham.edu.

Wendy Stubbs, Senior Lecturer in the School of Geography and Environmental Science at Monash University in Australia holds an MBA from the Wharton Business School and a PhD from Monash University. She is the Coordinator for the Master of Corporate Environmental and Sustainability Management program and teaches corporate sustainability in the MBA program. Her research interests include corporate social responsibility, corporate sustainability, and sustainable business models and systems sustainability. Her research explores new business models grounded in the principles of sustainability. Her research also focuses on systems sustainability, acknowledging that organizations can only be sustainable if the system they reside in is sustainable.

Charles Wankel, Associate Professor of Management at St. John's University, New York, earned his doctorate from New York University. Charles is on the Rotterdam School of Management Dissertation Committee and is Honorary Vice Rector of the Poznan University of Business. He has authored and edited about 30 books including the best-selling *Management,* 3rd ed. (Prentice-Hall, 1986), eight volumes in the IAP series Research in Management Education and Development, the *Handbook of 21st Century Management* (SAGE, 2008), and the *Encyclopedia of Business in Today's World* (SAGE, 2009), which received the American Library Association's Outstanding Business Reference Source Award. He is the leading founder and director of scholarly virtual communities for management professors, currently directing eight with thousands of participants in more than seventy nations. He has been a visiting professor in Lithuania at the Kaunas University of Technology (Fulbright Fellowship) and the University of Vilnius, (United Nations Development Program and Soros Foundation funding). E mail: wankelc@stjohns.edu.

Anatoly Zhuplev, Professor of International Business and Entrepreneurship at Loyola Marymount University, Los Angeles, received his PhD from the Moscow Management Institute, Russia. He taught for ten years at the Moscow Management Institute, and subsequently at the Advanced Training Institute of the State Committee for Printing and Publishing in Moscow; in Bonn, Germany in 1994, 1998, and 2009; in Warsaw, Poland (as a Fulbright scholar) in 2005; in Paris, France in 2004–2007, and at Northeastern University in Boston, MA in 1989–1990. He has about 80 books and articles on International Management, International Entrepreneurship, International Business, and Corporate Governance published in the United States, Canada, Western Europe, Russia, and the former USSR.

List of Figures

List of Tables

PART I

New Models for Globally Sustainable Businesses

CHAPTER 1

How Business, Resources, and People Fit Together in a Sustainability Model

James A. F. Stoner and Charles Wankel

Introduction

This book of chapters on global sustainability is based on one of the simplest of all premises: the premise that "what cannot continue forever *will not continue forever*." The way business operations are currently conducted, and the way other productive enterprises conduct their operations, cannot continue in their present forms because of the past and current rapid exhausting of the planet's natural capital—natural capital on which those operating ("business") models are built. It is hard to imagine that any, even slightly informed person could believe things will not change very dramatically in the very near future if for no other reason than the fact that we are using the planet's store of natural capital far more rapidly than it can be replenished by the cyclic processes that have maintained and restored it for millennia.

Because private for-profit business firms play such a major role in producing goods and services for society, they are obviously major users, and often abusers, of our limited stocks of the various forms of natural capital upon which our economic, social, and cultural well-being are based. Of course, business is not the only user and abuser—almost every institution and every one of us plays a part in this unsustainable (i.e., impossible to continue in this form) situation we currently find ourselves in. However, because business is such a big player in this game, it deserves special attention and has the opportunity to bring about the greatest level of change in our situation. Because the level of change that will be needed to move from our currently unsustainable world to a sustainable one is so great, it is not hyperbole to speak of the need for a global transformation of systems by which we produce and consume the goods and services that people around the world "enjoy" in varying degrees. Nor is it an exaggeration to speak of a "paradigm shift" that will be needed in how businesses see themselves in the world and how they operate in response to that vision.

The chapters in this book address some of the many, many ways in which the world is changing, and needs to change, as individuals and organizations seek to find more sustainable ways of being on the planet. The purpose of the authors and of the entire volume is to explore and share ways in which business organizations in particular can bring their operations more into alignment with the need for global sustainability.

PART I: New Models for Globally Sustainable Businesses

The three remaining chapters in Part I address three important aspects of finding ways for businesses to align themselves more fully with the imperative for a sustainable planet: getting for-profit companies to commit to conduct their operations in ways more consistent with the need for global sustainability; finding a brand new business model that will transform how businesses operate; and supporting new ventures that might test new models for producing society's goods and services in more sustainable ways. This broad task of transforming business in its most fundamental ways is the most challenging one facing the world, not just business organizations, today. Obviously the basic take-make-waste model on which the industrial system has been based since at least the dawn of the industrial revolution will not continue because the exhaustion of the world's natural capital on which it is based cannot go on indefinitely. Yet, the extent to which almost all of us have ignored the fact that "what cannot continue, will not continue" is impressive.

As a contribution to grappling with the question of how businesses can start acting on the imperative to align their structures and practices with the global sustainability imperative, Aileen Ionescu-Somers focuses on how the business case for sustainability can be sold by committed managers in for-profit companies, and the difficulties of trying to do so. For a number of years, she and her colleagues have been investigating the challenges of building commitment to aligning business practice with global sustainability needs in companies in general and in the food and beverage industry in particular. Chapter 2, "Business Logic for Sustainability," opens with a description of the global food and beverage sector that manufactures some of the world's most recognizable and widely consumed brands. The description of the industry shows how vulnerable it is to rapidly changing dynamics and megatrends that will ultimately greatly affect the industry's long-term financial sustainability. For a great many reasons, this industry is one in which the need to conduct itself in environmentally, socially, and politically sustainable ways is obvious and pressing, in part because the costs and dangers to it and to others of not doing are so severe.

In spite of these obviously very strong business reasons for discovering and taking action on the business case for sustainability, the results of several years of research carried out by the well-known IMD business school in Switzerland indicate that companies in the industry are making only very modest progress toward becoming more globally sustainable. They struggle, frequently unsuccessfully, to build the business logic for integrating social and environmental issues into business strategy. Building and rolling out business strategies focused on integrating complex sustainability issues presents difficult management challenges that are yet to be resolved in this industry, even if considerable efforts have already been embarked upon. Building on quantitative and qualitative empirical evidence collected from managers over several years and across several research projects, this chapter presents perspectives that can help business managers in formulating a more robust business case for sustainability in their organizations. The perspectives can be useful not only to managers in the food and beverage sector, but to managers in all industries.

Business cases for sustainability are sector-specific and can even be unit- or project-specific. Like any other business project, the logic for implementation of a sustainability strategy needs to be carefully built. However, the business logic for sustainability, to be robust, needs to capitalize fully on both tangible and intangible dimensions. Overemphasis on the tangible, quantifiable dimension will inevitably lead to a "short-changing" of the effort and thus the organizational strategy as a whole.

The chapter is also a veritable reality check against increasing hype around corporate social responsibility that may prompt consumers and the public to think that these issues are being fully addressed by companies. In fact, companies still struggle with these concepts and will be struggling for some time to come as long as sustainability issues are long term, as long as the business focus is increasingly based on short-term profits and outcomes, and as long as consumers do not "vote with their wallets" and insist upon sustainable products across the board.

In Chapter 3, "Sustainability as a Business Model," Wendy Stubbs addresses the need for fundamental transformation of the dominant take-make-waste business model of our times. In response to a growing recognition that business must move from "sustainability as an add-on" to "sustainability as a business model," Stubbs' chapter presents the "sustainability business model" (SBM)—a model where sustainability concepts shape the mission and driving force of the firm and its decision making. The SBM is an abstract model of a sustainable organization and was derived from in-depth case studies of two organizations, Interface and Bendigo Bank, that are leaders in implementing alternative sustainability-centric business models.

The first company, Interface, a global carpet manufacturer, is considered to be a leader in restructuring its business model around environmental sustainability. To achieve its sustainability vision to be "the first company that, by its deeds, shows the entire industrial world what sustainability is in all its dimensions," Interface developed a model of the "prototypical" company of the twenty-first century. This model embodies Interface's view of a sustainable enterprise: strongly service-oriented, resource-efficient, wasting nothing, solar-driven, cyclical rather than linear, and strongly connected to its constituencies (community, customers, and suppliers). The second company, Bendigo Bank, Australia's sixth-largest bank, is implementing a wealth-sharing community engagement model (CEM) that focuses on building "successful" communities. Community Bank is the first business based on the CEM. It is a branch-banking model that involves local people in solving their own banking needs. Generally, the community bank branch retains about 50 percent of the consumer products revenue it generates. Twenty percent of a community bank's profit can be distributed as dividends to its shareholders while 80 percent of the profit is set aside to fund further community development projects.

The SBM seeks to describe the structural characteristics and cultural capabilities of a sustainable organization. The research of Stubbs and her colleagues found that organizations can make significant advances toward achieving sustainability through their own internal capabilities, but ultimately organizations can only be consistent with the needs for global sustainability when the whole global system that they are part of is sustainable. That is, changes are required at the socioeconomic system level, both structural (such as redesigning transportation systems and taxation systems) and cultural (such as changes of attitudes toward consumption and economic growth). Nevertheless, organizations can make substantial progress toward sustainability by adopting the SBM as a design guideline.

In the final chapter of this section, Chapter 4, "Investing in Sustainable Entrepreneurship and Business Ventures in the Developing World: Key Lessons and Experiences of the EcoEnterprises Fund," Tammy Newmark and Jacob Park focus on fostering the creation and growth of organizations committed to creating new models for producing goods and services in ways that are consistent with the needs for global sustainability, and succeeding in making those new models work in the marketplace. With the overall goal of improving understanding of the theory and practice of investing in sustainable entrepreneurship and business ventures in emerging and developing economies, the chapter explores what we mean by and how well we understand the process and the institutional mechanisms behind investing in sustainable entrepreneurship and sustainable business ventures.

Newmark and Park note that many sustainability-committed businesses around the world never get off the ground because traditional sources of capital like banks tend to shy away from sectors that seem unfamiliar or too risky. This situation is unfortunate because new sustainability-oriented ventures can contribute to poverty alleviation and environmental conservation that are both critical to a more sustainable world, yet remain two of the largest challenges confronting the international community.

Increasingly, there is a call for the business sector to respond strategically to these issues by capitalizing on market opportunities posed by the pressures of poverty and environmental degradation in emerging and developing economies. This chapter is intended to assist businesses in heeding that call by presenting EcoEnterprises Fund, a fund committed to fostering such new sustainability-committed businesses. Their case study of the fund highlights important experiences from a decade of work advancing sustainable entrepreneurship and business ventures in emerging and developing economies.

Part II: Action Implications: Motivating and Influencing Business

The five chapters in Part II all address ways in which businesses can be enabled, encouraged, or perhaps pressured to move to more sustainable ways of making their contributions to the well-being of society and its peoples. That is, to fulfill their part of the social contract with the community that grants them the privilege of existing as legal entities.

The first two of those chapters, chapters 5 and 6, address the importance of providing businesses with the types of new employees who want to create a sustainable world and have the tools to contribute to doing so. Chapter 5 focuses on the daunting challenge of reforming our educational systems so they will provide the type of educational experience needed to enable graduates to contribute to a sustainable world. In Chapter 5 the focus is on bringing appropriate content on global sustainability into university curricula, and the difficulties of doing so. Chapter 6 addresses a successful endeavor to match the systems nature of the whole global sustainability *problematique* with education, training, and skills that are as multidisciplinary, transdisciplinary, and interdisciplinary as the systems problems we need to learn how to deal with.

Chapter 7 deals with how governmental regulators are emphasizing public disclosure methods to energize stakeholders to exert pressures on companies to improve their environmental performance. Chapter 8 focuses on the role corporate boards can play and how the establishment of a separate board committee focusing on environmental concerns might be a vehicle for bringing the actions of companies more into alignment with the need for global sustainability. Finally, Chapter 9 addresses the ways taxing systems can and are being used to encourage more sustainability-consistent behaviors.

In Chapter 5, "Sustainability in Business Curricula: Toward a Deeper Understanding and Implementation," Blanca Luisa Delgado-Márquez, J. Alberto Aragón-Correa, and Nuria Hurtado-Torres address the question of how and why different universities decide to integrate sustainable contents into their courses or choose not to do so. First they review the importance of providing university-level education for global sustainability in general and for business students in particular. Next they provide an international perspective on sustainable MBAs offered throughout the world and review the latest literature about the integration of sustainability into graduate and undergraduate business schools. They conclude the chapter with a report on a survey of Spanish business school decision makers influential in determining when, if, how soon, and how much content dealing with global sustainability issues and skills will be incorporated into the curricula of their schools.

Given the importance of management students acquiring sustainability-related skills to be used in their professional lives as managers, finding ways to facilitate incorporating such skills should be of interest not only to scholars who are analyzing the integration of new topics into business education but also to policy makers, to university administrators, and—most of all—to business leaders. Understanding the means by which formal and informal support can be provided to departments that are working to integrate sustainability issues into the curriculum can be useful to those committed to consolidating sustainability-focused education in universities and business schools.

Chapter 6 continues the theme of providing businesses with the types of contributors they need for the sustainability journey by focusing on an innovative program created by three French universities. In "The Contribution of Interdisciplinary Skills to the Sustainability of Business: When Artists, Engineers, and Managers Work Together to Serve Enterprises," Vera and Silvester Ivanaj discuss the ARTEM project, an innovative educational and institutional approach for combining disciplines in the service of education for global sustainability. This project is a unique venture in which art, technology, and management come together in an educational, research, and institutional project involving three prestigious academic institutions.

The authors argue that the junction of these three expertise domains promotes experimentation, creation, and implementation of innovative actions through the emergence of creative and innovative atmospheres and through continuous contact with professional environments. Ivanaj and Ivanaj address the links that exist between the interdisciplinary approach and the strategic and operational challenges that today's enterprises have to face when first committing to and then implementing a strategy focused on contribution to global sustainability. They describe how the content of the actions that businesses have to undertake is so complex that only an integrative approach using multidisciplinary, transdisciplinary, and interdisciplinary perspectives and skills will yield a proper understanding of, and ways of addressing, the current global sustainability challenges.

The authors also stress the important contribution educational establishments can play in the inter- and trans-disciplinary knowledge formation and skills development needed for sustainable development (SD), another very widely used term similar to the term global sustainability. The ARTEM-Nancy project illustrates the value of an educational philosophy that brings together engineers, artists, and managers to work on training initiatives in SD. The project is currently the only one of its kind, but the learning process it has adopted may be a useful one for other institutions to consider. The project creates independent learning through running projects in interdisciplinary groups, supported by professionals from companies and lecturers who bring their own very different contributions. These contributions are simultaneously artistic and creative, managerial and organizational, and technological or engineering. The process enables students to become instrumental in their own training through their choice among a variety of complex projects offered in a number of workshops.

Chapter 7 "Public Disclosure of Corporate Environmental Performance: Pollutant Release and Transfer Registers (PRTRs)" by Javier Delgado-Ceballos and Antonio Rueda-Manzanares turns to the important role public disclosure of corporate environmental performance can play in encouraging companies to align their operations more with the environmental needs for global sustainability. They describe how governments are moving beyond using decrees and market-based approaches as means of improving company environmental performance. In addition to these historically prominent approaches, governments are becoming increasingly interested in leveraging public disclosure of corporate performance, with a focus on PRTRs—Pollutant Release and Transfer Registers.

PRTRs require industrial installations to measure and report their emissions into water and air and onto land. The PRTR disclosures of installations' environmental emissions are documents available to the public and thus enable diverse sets of stakeholders to pressure companies into improving their environmental performance. The authors report that the PRTRs, such as the Toxics Release Inventory of the United States and the European PRTR, have been effective in leading companies to reduce emissions in a number of countries. Although they note that the information the PRTRs contain is much more valuable for comparative analyses than the environmental information published by the companies themselves, they conclude the chapter with suggestions about how the reports can be made still more effective.

Chapter 8 focuses on companies' board of directors—one part of a company that can play a major role in initiating and supporting a corporate-wide commitment to global sustainability. Although the role could be large, board leadership in this domain seems to have been neither frequent nor highly visible. In Chapter 8, "The Adjustment of Corporate Governance Structures for Global Sustainability," Natalia Ortiz-de-Mandojana, J. Alberto Aragón-Correa, and Javier Aguilera-Caracuel contrast companies that do and do not have separate board environmental committees. They start with the very reasonable premise that for companies to move to models of sustainability, the adjustment of the corporate governance structures to climate change concerns can be an important step. One way firms can adapt their corporate governance structures to a more environmentally proactive posture is by assigning responsibility for environmental matters such as climate change to a specific committee. They reason that a specific committee responsible for environmental issues might provide support to managers regarding environmental issues, ensure the organization's compliance with all aspects of environmental legislation, and raise the attention paid to environmental matters.

Using a sample of 707 firms from 21 different countries based in North America and western Europe, they found that the type of industry, country, and size of firm were related to the likelihood that companies would have separate environmental committees (that they would "delegate environmental issues to specific committees" to use the authors' phrasing). In their study, larger firms and companies in highly polluting industries and in regions with more stringent environmental institutions were all more likely to have such separate board committees. Although the researchers found that companies with board environmental committees were more likely to publish public information about how they faced climate change concerns, they were not found to be more likely to develop new environmentally friendly products.

Chapter 9 discusses a topic surely near and dear to the heart of all executives: *taxes*. In "Taxes and Sustainability," John Stancil examines the history and success of measures that have been utilized to encourage actions consistent with a more sustainable world and to discourage ones inconsistent with global sustainability. He observes that taxes have long been used in society to encourage or discourage certain behaviors. If something is seen as desirable, oftentimes a tax credit can be offered to encourage the desired behavior. For those actions seen as undesirable, the government can levy taxes to discourage the behavior. These can take the form of taxes, fines, penalties, or user fees. In recent years, the use of taxes as a tool for social purposes has increased significantly.

A number of approaches have been tried to "encourage" businesses to act in a more sustainable manner. Although past measures have had some success, their approach has been piecemeal. Stancil argues that sustainable tax measures will succeed only on a global basis, to prohibit firms from moving to areas that are less aggressive in promoting sustainability. Seven factors are presented that are necessary for a successful global sustainable tax policy—neutrality, comprehensiveness, coordination, a Pigovian approach, removal of subsidies, social equity, and visibility. To close the chapter, he discusses tax policy as a part of an overall sustainable economic policy.

Part III: Business Transformations of and by Its Environment

The six chapters in Part III address examples of how businesses are being, or will be, influenced by their environments and how they, in turn, are influencing their environments. Chapter 10 provides an overview of the International Energy Agency (IEA) and the challenges it has taken on. Chapter 11 discusses how a stereotypically nonsustainable industry, the plastics industry, can and is finding ways to become more sustainable. Chapter 12 explores the role of waste-pickers in some Latin American countries, how they are contributing to effective recycling and how they can be supported in doing so. Chapter 13 explains how global forces are encouraging wineries to adopt sustainable practices. Chapter 14 considers how farmers' markets (FMs) contribute to localization and shortening of supply chains and the benefits that accrue from their doing so. Chapter 15 addresses ways stakeholders can combine their efforts to develop new insights into sustainability options.

In Chapter 10, "Why Changing the Way We Use Energy Is Essential for Global Sustainability," Richard H. Jones provides an overview of the IEA's emerging endeavors to reduce the world's dependency on fossil fuels. He describes the IEA's founding and development of an emergency response system, which has been fully activated twice, after both the Iraqi invasion of Kuwait and the 2005 hurricane damage to oil production infrastructure in the Gulf of Mexico. He discusses new emerging challenges that transcend oil and involve assuring wide access to all forms of energy including natural gas and electricity. The trend of non-OECD countries to move ahead of the OECD countries as the largest consumers of energy and the challenges of reducing the environmental impact of energy use are discussed. He assesses the world energy outlook and discusses the consequences of staying on the current path, given the rapid growth of global greenhouse gas emissions. Other topics discussed are the consequences of climate change and the 450 scenario for achieving the necessary abatement. End use efficiencies and the decarbonization of the power and the transport sector are covered. He then assesses the cost of moving to a low-carbon economy, including the benefits of doing so. Financing the transition is a great issue for developing countries that look for help from the more developed ones. However, the level of support mechanisms for providing it and how the relative burden would be shared are matters for negotiation. The significant costs for delaying action on climate change are assessed. Many of these costs are associated with the inherent inertia of the energy sector. The chapter concludes with a consideration of the world's situation after the Copenhagen Accord and the continuing role the IEA can play. Because of the critical importance of effective communication between policy makers, stakeholders, industry, and the public, the IEA is searching for audiences for its analyses beyond its traditional one of government officials and policy makers in member countries. In doing so, it is using new social media such as Facebook to spread its messages, as well as streaming media on energy efficiency on its Website.

Avrath Chadha of the Swiss firm Hoerbiger focuses on evolving changes in the polymer plastics industry in Chapter 11, "From Carbon to Carbohydrates: Toward Sustainability in the Plastics Industry." With a global polymer production of 250 million tons, the plastics industry is one of the chemical industries' major sectors. The plastics industry has historically had incremental process-oriented innovations rather than large-scale product or process innovations. However, radical change in the plastics industry is inevitable since polymers produced from fossil fuel raw materials are tied to oil reserves that will become increasingly expensive as they continue to be depleted.

A green chemistry orientation emphasizing renewable resources is underway. The current focus is on biopolymers or bioplastic processed from renewable feedstock like starch. Some of the end products are biodegradable. This chapter uses case studies to explore the different reasons firms have for entering the biopolymer market and emphasizes the various strategies used for biopolymer product development. New technologies in plant breeding and processing are narrowing the cost differential between biopolymers and petrol-based polymers. Advances in biotechnology, particularly in genetic engineering, will play an important role in further development of biopolymers.

Some firms report that customer queries about more sustainable product solutions sparked company innovation efforts. The activities of competitors have also played important roles. A number of firms allow bootleg research where scientists can start informal research on this technology without having to go through their bureaucracies. Some firms have developed close collaborations with customers and suppliers in the form of R&D consortia to increase their knowledge of future product requirements. The industry seems to be in an epoch of exploration, experimenting with different technological and product alternatives. Decentralized cross-functional project teams are helping to develop successful biopolymer product technologies. The replacement of petrol-based polymers with biopolymers or other substitutes is imperative for a movement to a more sustainable world. Although products made of biopolymers are still niche applications and large investments in research and development with inherent financial risk are necessary, Chadha concludes that they are certainly the way to the future.

In Chapter 12, "Informal Waste-Pickers in Latin America: Sustainable and Equitable Solutions in the Dumps," Candace Martinez examines the institutional changes that the informal recycling sector in Latin America has been experiencing in recent years and what the social, economic, and environmental effects have been. Governments in several countries (Brazil, Colombia, Peru, and Argentina) have legitimized the work of waste-pickers or informal recyclers by referring to them legally as "entrepreneurs." In theory, such official status allows waste-pickers to bid for municipal waste-management contracts; in practice, unless waste-pickers are organized in collectives and/or associations, it is difficult for them to compete with established firms. Nevertheless, informal recycling associations in Latin America have made great strides in improving the livelihood of their members and made important contributions to the sustainability objectives at the country, city, and firm levels. The chapter begins with an overview of the solid waste and recycling sector in the developing world and the challenges that informal recyclers face. Martinez then draws on new work in institutional economics to illustrate that the institutional changes that the sector is undergoing have historical antecedents and to note that the rate of change for formal institutions is different from that of informal ones. A formal institutional change (rules, laws, regulations) can occur as soon as the ink dries on a piece of legislation, while an informal institution (beliefs and attitudes) is gradually absorbed by a society. Therefore, theory would predict a longer time horizon for potential societal acceptance of the traditionally scorned-upon and marginalized informal recyclers. Evidence from Latin America suggests that the informal is converging with the formal. The bulk of the chapter addresses the successes of informal recycling associations in the region and how their expertise, experience, and qualifications in recycling have managed to win the support of national governments and the private sector. The outcomes are synergistic government, civil society, and business partnerships that showcase what participatory solid waste management can achieve: a solution to the world's trash problem that is socially acceptable, environmentally sustainable, and economically sound—a *gana-gana* for all stakeholders.

Anatoly Zhuplev and Cynthia Aguirre in Chapter 13, "Sustainability of the Wine Industry: New Zealand, Australia, and the United States," present the findings of a comparative study of sustainability trends in the wine industry, particularly in the New Zealand, Australian, and Californian markets. The study is based on extensive library research and 15 field interviews with the wine industry associations, trade organizations, wine makers, marketing professionals, and other experts. Similarities and differences in attitudes and practices toward sustainability are explored across the three countries.

The 15 wineries and organizations in the wine industry included in the study have adopted their own unique approaches toward sustainable practices. However, there are general sustainability trends and goals emerging industry-wide. Consumer demand for sustainable products and the changing global economy are leading many wineries to shift their focus toward sustainability-oriented practices. Although each organization may have different motivations for engaging in sustainable practices, there is a consistent theme of personal values and a consciousness about the environmental impact that the wine industry makes throughout New Zealand, Australia, and California. Companies receiving organized support from government agencies, winery associations, or simply a sharing of ideas were found to adopt sustainable practices more readily. In New Zealand, Australia, and California, the wine industry members are making a significant effort to reduce their environmental impact and preserve their land for future generations.

In Chapter 14, "New Trends for Sustainable Consumption: The Farmers' Market as a Business Imperative for the Reeducation of Consumers," Alessio Cavicchi and Benedetto Rocchi underline the role of Farmers' Markets (FMs) in reeducating consumers to the value of rural sustainability. In doing so, they describe the role of FMs in bringing consumers back to the sustainable values that held sway in previous times. FMs around the world are put forth as a key response to the unsustainability of conventional food production systems. FMs probably are the oldest common type of direct marketing. Bringing food producers and consumers into close contact is a paradigmatic example of an *alternative food network*. Given their characteristics, FMs are at one end in the continuum of forms that a *short food supply chain* can assume, according to their extension in space and time. In the past two decades, they have represented the first experience in relocalized and resocialized forms of exchange for a growing number of consumers. The primary motivations evoked by consumers when interviewed about their willingness to make purchases from FMs are: better food quality, locally produced foods, higher social interaction potential, and learning directly about the vendors and their food production practices.

Denise Barros de Azevedo, Eugênio Ávila Pedrozo, and Guilherme Cunha Malafaia's Chapter 15, "Participation of Agribusiness Stakeholders in Global Environmental Questions," discusses climate change, its influence on bioenergy developments in Brazil, and the ways participation of stakeholders in environmental debates has fostered the discussion of new themes locally, nationally, and globally. Global climate change trends reflect the behavior of individuals, so working to achieve a greater understanding of those trends and associated issues is important. The authors discuss initiating dialogues among stakeholders from different fields of knowledge on climate change and agribusiness. Globally, industrial progress has been dependent on the development and use of nonrenewable fossil-fuel-based energy sources, but the increasing scarcity of those fuels is raising their costs and the impacts of using them are overwhelming the capacity of the environment to absorb the CO_2 they release when burned. Continued industrial progress will be premised to an increasing extent on the discovery of new energy sources. The search for bioenergy to address climate change becomes a strategy for the development of organizations, their stakeholders, and society. The huge participation of nonrenewable sources in the world energy supply challenges society to focus on the search for alternative sources of energy. Although bioenergy offers opportunities for substituting for nonrenewable energy sources, its development creates risks of increasing food security. Impacts on food security can vary depending on the evolution of the market forces and technological developments, both influenced by political choices at national and international levels. The authors describe how dialogue among stakeholders favors finding better alternatives and solutions to conflicts and clarifies the usefulness of the natural environment.

The authors in this volume emphasize that we are entering an epoch where businesses must eschew a myopic focus on short-term profits and look to the burgeoning opportunities associated with achieving global sustainability. They suggest a number of promising ways businesses can move forward in finding and creating such opportunities. Taking actions to seize those and many other opportunities is increasingly demanded by customers, governments, and the competition.

CHAPTER 2

Business Logic for Sustainability

Aileen M. Ionescu-Somers

Business thought leaders have recently been raising more urgent questions about corporate responsibility in the aftermath of the currently ongoing financial crisis provoked by "bubble excesses" and other high profile business misdemeanors. Some commentators have suggested that the dramatic events we have witnessed and their serious consequences for the well-being of populations worldwide provide fertile ground for a new breed of responsible company.

However, research carried out by the Center for Corporate Sustainability Management (Center for CSM)[1]—a learning center at IMD, a leading business school—indicates that such a development is far from certain. While, precrisis, many global companies had already begun to tune into the concept of "doing well by doing good" and linking their business activities to corporate sustainability, corporate social responsibility (CSR), or whatever other term they give to the idea of integrating economically relevant social and environmental issues into corporate strategies. Let us not forget that this challenge has been around for a relatively long time already.

The economic theory of "externalities" explains why corporations in general are inextricably linked to social and political contexts in ways that go beyond the obvious market transactions (Steger 1998, p. 14). The concept has been around since 1923 and applies to both positive and negative effects of corporate activities outside of market-valued costs. Examples of positive externalities are the familiar income multiplier of local investment or the synergies of agglomeration, while local pollution is an obvious forerunner as a visible negative externality since it is frequently highly attributable to a specific firm's activities. Although companies have increasingly been pressured to take responsibility for their negative externalities, the battle is far from won as yet. An extreme example is the failure thus far of oil majors to resolve pollution and social issues exemplified by the catastrophic oil spill currently being played out in the Gulf of Mexico and in the Niger Delta (Ionescu-Somers 2006).

More recently, highly complex negative externalities such as climate change or social issues such as child labor—sometimes much less directly attributable to the actual operations of specific companies—have been projected onto the corporate radar screen by stakeholders. As a result, already today, few high profile global companies ignore the need to address such issues more systematically. Sometimes addressing these issues offers an opportunity to go beyond traditional philanthropic responses and opens up possibilities that make real business sense. At the very least, companies that take a systematic approach to these issues may be able to escape the sharp ire of activists and media, thus doing a better job of managing an increasingly complex stakeholder environment. But the strongest case for taking action occurs, of course, when managers discover that by being increasingly sustainable from an environmental and social standpoint, firms can also increase financial sustainability and deliver shareholder value.

The Center for CSM at IMD has been investigating the evolution of business logic for sustainability in various industries since 2002, prompted by multiple interactions with its members and other stakeholder groups. A comment from one senior executive typifies the challenge faced by many managers:

> Our senior management takes a compliance-driven view of environment, health, and safety. This creates a problem when trying to deal with sustainability and corporate social responsibility. They are not against, they just do not understand. I need to defend everything that my group does outside of compliance and provide convincing information about the contribution to shareholder value of "sustainable" management?

If this conundrum sounds familiar to many readers, that is not surprising; managers across the world face a similar dilemma. Finding the logic for integrating sustainability into business strategy in the area beyond compliance is complex and challenging. Case-makers often face a highly skeptical management with a "what's in it for us?" stance in relevant business units, and demands for definitive quantified proof that sustainability positively impacts the corporate bottom line. Understandably, such business units are pinned to the wall with shareholder demands for short-term bottom line outputs, and particularly so during a financial crisis such as the one we are currently experiencing. So, longer-term issues can end up being placed "on the back burner." With critical issues such as climate change, this situation has resulted in a serious global dilemma.

Thus, we carried comprehensive studies of industry-specific business cases for sustainability, the roll-out of related strategies, stakeholder perceptions of corporate sustainability performance, and effectiveness of sustainability partnerships (Steger 2004; Steger 2006; Ionescu-Somers et al. 2008). We not only used extensive desk research, but also carried out hundreds of surveys and interviews of business and sustainability managers across multiple industry sectors and stakeholder groups, convinced that perception represents reality with such value-loaded concepts (put bluntly, if managers are not convinced of the value of a proposition, then it is highly unlikely that they will act upon it).

We discovered that seeking the business logic for sustainability entails relating sustainability projects to what we called the "Smart Zone" (see Fig. 2.1), the space in which the company creates additional economic value by improving its environmental and social performance beyond that required by governments through legislation. Note that, in any case, since countries vary greatly in their level of internalization of externalities through legal instruments, so too does the business logic for sustainability for companies in different countries. In the figure below, the inverse U-shaped curve describes the relationship between corporate economic and environmental and social performance. Zone 1 represents the "Smart Zone," where the most profitable projects are identified; the curve is therefore steep. In Zone 2, the Net Present Value (the revenue stream discounted with the risk-weighted cost of capital, or the internal hurdle rate) becomes negative, with overall economic performance declining. In this zone, one can legitimately suppose that for various reasons (reputation, licence to operate, brand value) companies may still choose to allocate resources to subprofitable projects while not actually losing money. However, there comes a point when they will reach Zone 3, where environmental and social performance is no longer adding economic value and therefore the company is effectively losing money with the investment.

The reader should note that the diagram is intended to be conceptual rather than represent a mathematical or quantitative model against which to measure the business logic for sustainability; it helps managers to understand and translate into interpretations of where a project they might be considering for implementation is positioned on the curve.

In the process of exploring how managers find the "smart zone," we came up with "building blocks" for the business case for sustainability based primarily on managers' own perceptions. We evaluated the extent to which managers are convinced that by integrating sustainability issues into business

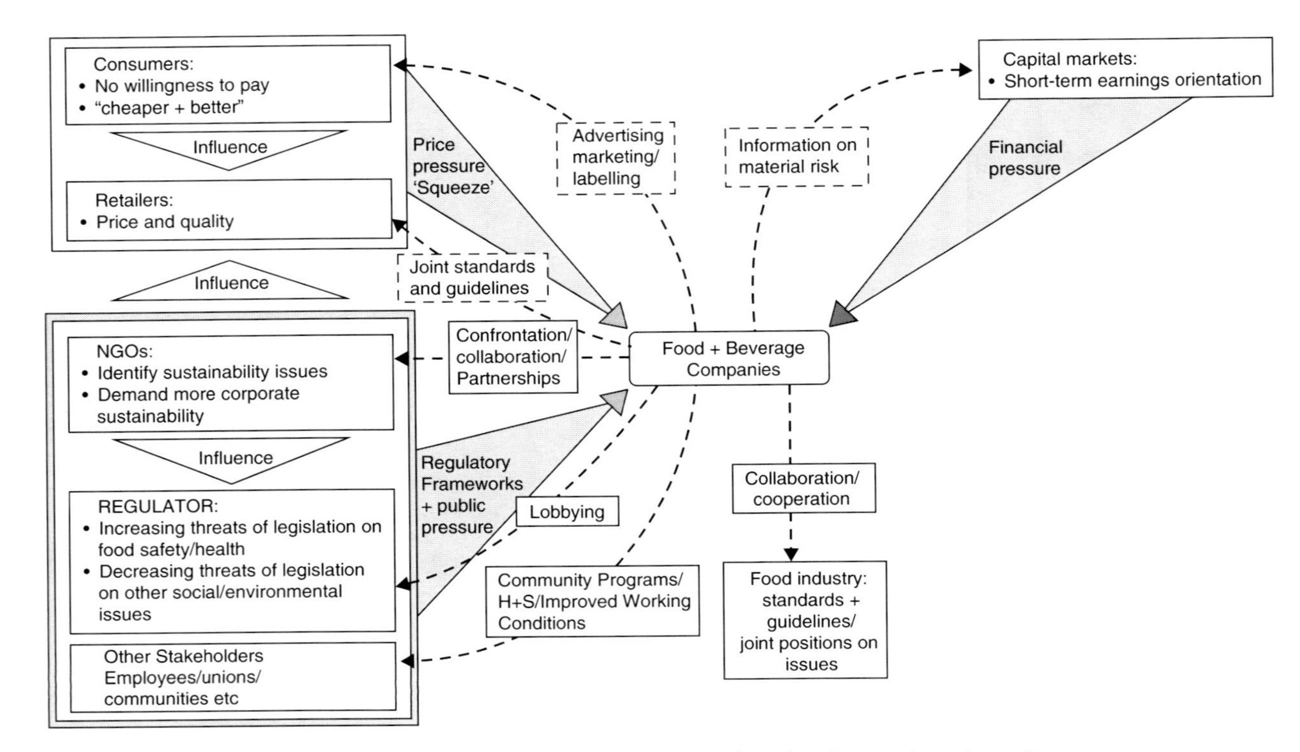

Figure 2.1 The relationship between economic performance and environmental and social performance beyond compliance (Ionescu-Somers 2008)

strategy and focusing on the business opportunities they bring, companies can better create economic value. We assessed the strength of business cases and their application across industries, but also looked at promoting factors and barriers that do or do not enable industries to align behind sustainability strategies.

The Food and Beverage Industry Example

Since our early research (Steger 2004) was primarily concerned with establishing a *sector-specific* business case, we back up some of our arguments in the following sections with findings from food and beverage (F&B) manufacturing industry research (Ionescu-Somers et al. 2008).

Linking Sustainability to the Competitive Environment

Although there are a myriad of small and medium-sized enterprises in the F&B industry, we focused on the global manufacturing industry led primarily by highly branded multinational corporations such as Nestlé, Kraft, Danone, and Unilever. These F&B companies are fiercely competitive and thus seek sophisticated strategic options to maintain coherence and growth, along with new sources of advantage allowing strategic renewal. Further competition from retailers' own brands leads to a quest for product differentiation. The ultimate objective is to gain enough coveted retail shelf space to ensure high levels of consumer brand loyalty to remain profitable.

Globalization has broken down boundaries, thus increasing competition and firm rivalry. Markets, technologies, regulation regimes, consumer behavior, and geopolitics are now subject to an array of discontinuities and global megatrends that create continuous flux in the competitive relationships within the F&B industry. And unlike in some other industries, F&B companies are highly visible to the consumer, who typically regards food from an intensely emotional and personal point of view.

Competitive advantage can be achieved in two ways: by lowering prices or by providing superior value through, for example, process and product improvements. Lowering prices is ultimately not in the interests of a sustainable agenda since it provokes a counterproductive "race to the bottom," in which one company's efforts to achieve a more favorable outcome at the expense of other companies causes retaliation, eventually resulting in an overall less favorable outcome for all companies. In seeking profitability, leading companies are following the second strategy, for example, by moving away from commodity products—animal husbandry or crop production have low profit margins and thus low share prices—and vertical integration.

Instead, they are seeking supply value-added (often based on the brand) allowing them to increase prices. F&B industry products dominating the market are highly processed and brand-labeled. Lucrative profits are reaped from value-added industrial-scale production and distribution. But, can sustainability also be part of the value-added equation?

To answer this question, it is interesting to look more closely at global megatrends. Scientific studies carried out by the World Wildlife Fund (WWF) conclude that current consumption is already 25 percent over the planet's capacity to regenerate resources to support the global population. And global population is expected to increase up to some 10 billion by 2050, some 50 percent more than currently. This means the pressure on finite resources will increase exponentially. Agricultural systems, upon which the F&B industry heavily relies, already occupy more than half the habitable areas on the planet. Water scarcity is prevalent and manifesting itself—owing to multiple factors such as climate change and poor land use—in many regions around the globe, and agriculture consumes some 70 percent of available water resources every year (de Fraiture and Wichelns 2010). Manufacturing companies, retail outlets, and food service environments have reacted and adjusted to major lifestyle shifts in developing countries: increased consumption of food in restaurants (including "fast food"), more women in the work place and single parent families (with less time to dedicate to cooking at home), more food "grazing" rather than fixed family meals, and so on. Food service establishments are getting an increasing "share of stomach" in the United States and Europe.

Emerging economies, source of future growth for the industry, are following suit. Many are also undergoing substantial changes in dietary habits as better-off populations move from a vegetarian to meat-based diet, such as China. The UN's Food and Agriculture Organization estimates that meat production produces one-fifth of the world's greenhouse gas emissions; yet, to meet demand for meat given demographic growth and the current escalating rates of consumption, production would have to be doubled by the year 2050. To compound the problem, biofuel production is increasingly commanding cereals that were heretofore utilized for other food purposes. This has led to shortfalls in the global marketplace, food price increases, and even food riots in some developing countries, with considerable ramifications for sustainable development.

Prevalent market forces affecting the pace of sustainability activities in companies are the weakness of fragmented suppliers. Many thousands of poor farmers in developing countries, for example, supply the bulk of commodities to the industry and yet, individually, have little clout. The retailer buyers, themselves trapped in a cutthroat business environment, exert a significant hold over the food system. The manufacturing sector is "sandwiched," as it were, between these two countervailing forces.

We took a snapshot of current F&B sustainability issues, as perceived by managers, and related them to their economic relevance. This allowed us to (1) achieve a more holistic picture of external forces, other than purely market factors, which affect the industry; and (2) understand how these external forces translate into sustainability issues for companies.

By managers' own admission, our current food system is far from "sustainable," or even one that assures current and future societal and environmental stewardship. Upstream, because of highly intensive farming of raw materials, the industry contributes to global warming, overfishing, deforestation, biodiversity loss, soil erosion, pesticide build-up, and water shortages. Furthermore, there is concern about industrial activities creating unfair trading practices or human rights, slave, or child labor issues. Downstream, there are major health concerns related to obesity, alcohol abuse, long-term health effects of highly processed foods, and the issue of consumer choice. Companies declare with some legitimacy that, although the industry influences consumers through advertising, there are also lifestyle changes and work pattern fluctuations outside the control of the industry that also influence eating habits of populations, also contributing to the rocketing obesity statistics and related health problems affecting not only United States and Europe but emerging economies also.

Nevertheless, such issues provide fertile ground for food industry critics to undermine consumer confidence in the F&B beverage sector, as well as provoke processes leading to new layers of regulations. Given their potential to strategically alter the direction of future corporate operations (the threats to raw material supply and the ominous equation "no supply = no business" is an obvious one), we may easily conclude that such economically relevant sustainability issues cannot simply be ignored by food companies. But the reality we found is that companies are challenged to develop a coherent sustainability agenda given the diversity, fragmentation, and sheer number of value-laden, intangible, and relatively "fuzzy" issues that hit the corporate radar screen.

Business managers are pragmatic people, often confused about what sustainability issues are and how to prioritize them. Particularly when it comes to issues upstream or downstream of operations, they do not necessarily see what competitive advantage addressing them brings their company, or how they might apply to daily business operations. Unless an issue can be grasped quickly using the manager's own language, discussions about sustainability may appear to have little relevance to their myriad ongoing tasks. Since managers have a much clearer obligation to meet the needs of their shareholders and customers than those of their other stakeholders, a measured corporate strategic response to sustainability issues outside of regulation poses a problem. The result is somewhat pessimistic: sustainability issues can fester until—under pressure from NGOs and the media—the defensive/reactive mode is called for.

It is thus important for managers promoting sustainability in organizations to build a robust business case for "killing three birds with one stone" by focusing more fully on the creation of economic value through the improvement of environmental and social performance. Managers know well that economic decisions always result in more than merely economic effects. However, predicting all positive and negative consequences of decisions is difficult and not always practical. Complexity and a volatile business environment make it even more difficult. This is particularly true in these troubled times.

Leading global companies are using a materiality approach to identify sustainability issues that are of specific economic relevance to themselves and their sectors. They view addressing such issues as a competitive necessity. For example, in the F&B industry, consolidating a competitive position by ensuring that raw material supply is secure seems like a clear starter; yet the challenge is not easy:

> Policies currently don't allow externalities to be reflected in the cost of raw materials, but by doing so, the business case for sustainability in the supply chain would be greatly enhanced.
>
> Senior business manager, Manufacturing and Supply chain

Nestlé, Unilever, and Danone created the Sustainable Agriculture Initiative,[2] a membership platform for the F&B industry to jointly agree upon sustainable agriculture guidelines and standards that can be linked to sourcing and procurement policies. Another partnership, the more U.S.-based Sustainable Food Lab, also has this issue foremost on its agenda.[3] Under-resourced and with a plethora of challenges, these initiatives have not yet succeeded in forcing the industry's hand to take a giant step (Steger et al. 2009). The economic framework simply does not yet lend itself to such action.

However, as Reinhardt points out, the concepts of sustainable development and corporate competitive advantage are not entirely dissimilar:

> When reduced to its economic essentials, the Brundtland injunctions not to impoverish the future to enrich the present boils down to an injunction to maintain the total stock of capital at its current per capita levels.
>
> Reinhardt 1999

Reinhardt argues that the notion of Triple Bottom Line sustainability is a business idea closely linked to the idea of sustainability in the true "business" definition, that is, "the fundamental basis of above-average (financial) performance in the long run is sustainable competitive advantage." And, as Porter underlines, "without a sustainable competitive advantage, above-average performance is usually a sign of harvesting" (Porter 1985, p. 11).

To a significant extent because of a focus on short-term profitability, paradoxically, the long-term financial sustainability of the industry is at risk. And due to the lack of a "level playing field" (Ionescu-Somers 2006), no one company is willing to jeopardize its own competitive position relative to others in the short- to medium-term.

Risks and opportunities emanating from sustainability issues can emerge from any aspect of a company's business; its internal operations, suppliers, or customers; or even from unpredictable factors such as climate. If companies are not proactive on such issues, they not only run business risks but they are not grasping opportunities to deliver shareholder value. They will not be prepared to respond to consumers and activist groups on questions related to social and environmental responsibility. Consumer trust in their brands will be eroded, and reputation threatened. Efficiency may be lost, and regulatory burdens increased. Add to this the loss of potential first-mover advantage and one might say that there is an "open-and-shut" business case. However, as a senior business manager in communications commented:

> There *is* an 'open-and-shut' case—but companies don't see it. We benchmarked with our peers. Few companies are really taking it on board. The sector, highly conservative, is slow to come to the table.

Stakeholders: An "Uncertain Trumpet"

So F&B corporations are being pushed by their stakeholders to take responsibility for a host of negative consequences of economic decisions occurring as a result of actions either much further up or down the food value chain. Accountability for companies is toughening, challenging the mental models of traditional business as a result. F&B companies nowadays have to keep ever a closer eye "outside the factory gate" (Billington et al. 2009). Also, changes have occurred within an exceedingly short time span: whereas in 1993, a leading journalist wrote an overview of key trends relevant to the food industry and did not mention a single associated sustainability issue (Giles 1993), such an analysis would be entirely different today.

Integrating sustainability issues, we found, often depends on the level of pressure companies receive from stakeholders to recognize these as risks and grasp emerging opportunities. Without external pressure, in light of the complexities they are already dealing with on a daily basis and with short-term need for profit, business managers tend to shun the possibility of making their lives more complicated in the interim by taking on a set of perceived "fuzzy" issues.

But to ensure true sustainability of the food value chain, all stakeholders need to increase awareness of what comes before and what comes after their contribution, with every link in the chain delivering value to the end-consumer. To do this, companies need to work at a global level with other actors in the food system such as governments, farmers, producers, distributors, retailers, consumers, scientists, and so on. This involves consensus-building, not a common industry approach, which can be slow and sometimes painful to managers.

We evaluated the push-pull relationships between those stakeholders that promote a sustainability agenda and those on the other side that deter decision makers from taking faster action. This helped us to understand the factors prompting stakeholders to action and informed our understanding of why companies are not moving toward "strategically integrated sustainability" more rapidly.

Unsurprisingly, given their influence on economic sustainability of the company, stakeholders from the business sphere are still "calling the shots." The most critical countervailing factor is the consumers' unwillingness to pay extra, mainly because they do not perceive individual personal benefits through doing so. Identifying, describing, and conveying consumer personal benefit through sustainable purchasing requires some attention from companies before any marked change can come about.

In spite of increasing managerial awareness about the relatively volatile stakeholder environment in which they are operating, nonbusiness stakeholders, such as NGOs, still remain unfortunately less relevant to managers (Prinzhorn et al. 2006). Yet given the need for global F&B companies to protect their brands, the prospect of NGO activism is more likely to have an impact on highly branded companies than on those less exposed. Moreover, F&B managers interviewed told us that to base company strategy on a stakeholder model gives a new dimension and economic rationale to a business strategy that would otherwise be overthrown, promoting organizational learning and mitigating pressure, but also protecting reputation and brand value:

> If our company had not built up its network of stakeholders in the comprehensive way that it has, it would be insular, inflexible, and unable to make decisions as well as today.
>
> Senior business manager, Strategy

IMD carried out an extensive study of the stakeholder environment in 2006 (Steger 2006) to look at perceptions of nine stakeholder groups of

CSM. We analyzed the level of self-perceived leveraging potential by each group on prompting business and industry in Europe to more accelerated action. Many of the stakeholders, including NGOs, ranked themselves low in terms of their own effectiveness. We also found that despite the "hype" behind stakeholder pressure in both the academic and business community today, stakeholder pressure is decreasing somewhat in Europe as companies become adept at managing their stakeholder environment. NGOs feel "managed" and managers do not seem to be "feeling the heat" (Ionescu-Somers 2006; Prinzhorn et al. 2006), as "more bang for the buck" is being sought by NGOs in the emerging economies.

The Value of Value Drivers

Figure 2.2 represents the current state of stakeholder pressure and response dynamics in the food processing industry, with strong retailer "squeeze" (augmented by strong consumer pressure) for ever lower prices and better quality, shareholders that pressure the industry for lucrative short-term returns, and an array of other stakeholders that are not exercising enough countervailing influence on the "beyond compliance" sustainability agenda to reverse the trend of strong industry response to the price/quality demands of retailers and consumers. Regulators have a strong role in promoting CSM through threats of further regulation. The industry spends a lot of time trying to stave off future potential impact such that it retains global competitiveness. Note that in the figure, we have included not only the type of pressure on the industry, but also the typical industry response; the broken-lined boxes correspond to the weakest areas of response. We identified weak responses in labelling/advertising and marketing efforts, cooperation with retailers on standards and guidelines, and information on material risk provided to capital markets. The strongest are the responses to public and regulatory pressure and intraindustry collaboration. Given shareholder pressure for short-term earnings, companies still hesitate to make sustainability-focused investments that are not clearly linked to near-term cash flow.

Radical innovation for new business products and processes is also weak, while more incremental innovations, based on eco-efficiencies and step-by-step improvements in existing business models, are currently the stronger industry response. But by interpreting shareholder value too narrowly, companies can undervalue or ignore the value-driving potential of sustainability. Indeed, excessive focus on the short term could mean that companies are missing important growth prospects. The industry's rather weak reaction to the ongoing seriousness of the obesity issue demonstrates the danger of

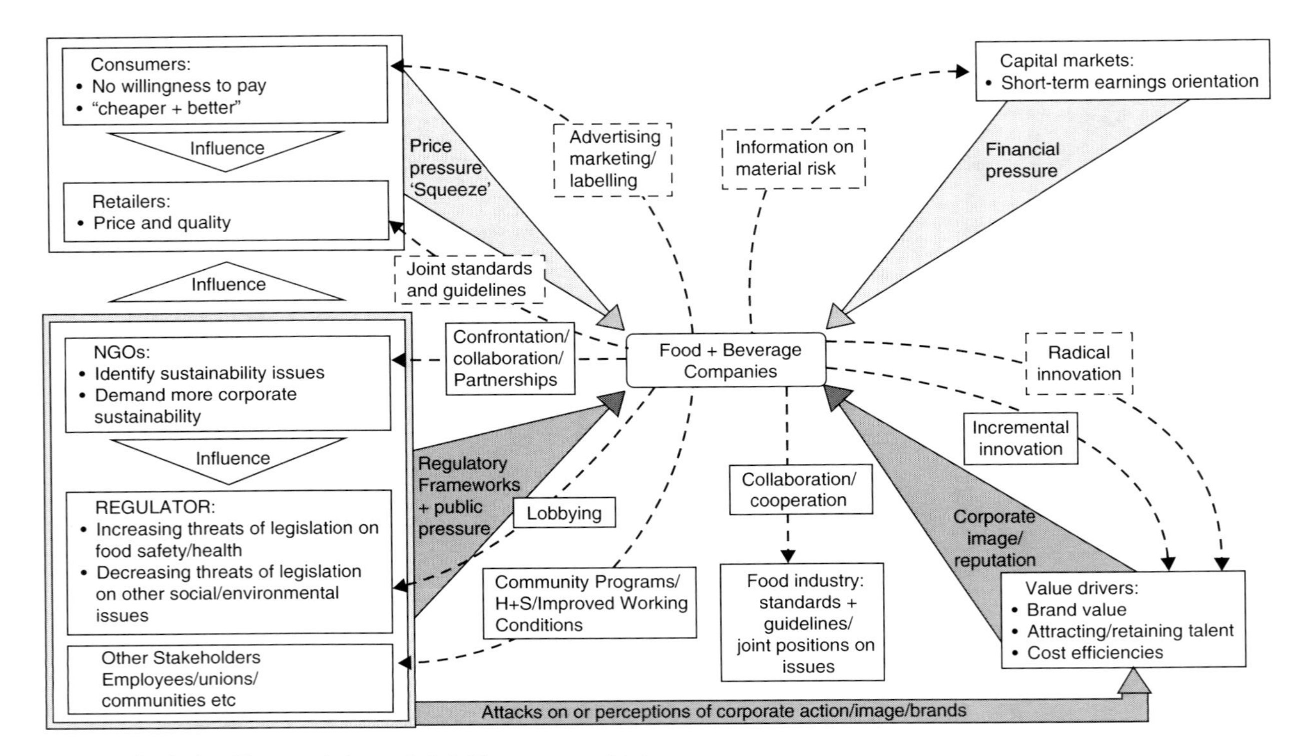

Figure 2.2 The food and beverage industry: Stakeholder pressure and industry response

focusing exclusively on short-term business performance and neglecting longer-term opportunities. Many more new or revised product offers are required in F&B companies today to address this escalating issue.

We suggest that the problem goes way beyond the sustainability agenda and touches the domain of industry renewal in general. Given its "front-line" exposure, the industry has an opportunity to evolve from a defensive, vulnerable position to a proactive and involved position as a legitimate member of the society in which it operates. Given where its future growing markets are likely to be even in the midterm (emerging economies), it is in its business interest to do better. Pioneering companies such as Unilever are experimenting with radical innovation, but a risk-averse "softly softly" approach predominates the industry, mainly due to the aforementioned cutthroat business environment. Proactive companies have to manage an already existing, strong strategic position while moving slowly toward a new paradigm.

The contributions of CSM to business value drivers are underexploited. Leaders believe that the positive impacts of integrating corporate sustainability strategies will be increasingly recognized and rewarded through brand trust, employee loyalty, and reputation benefits. Note that the greatest challenges in this area do not relate to what we call the "low-hanging fruits," such as the efforts of managers to reduce operational risk and introduce measures to increase eco-efficiency. Those aspects are relatively easily measured and easily managed—and are also tantamount to good management. It is the more intangible positive relationships between environmental/social performance and financial performance within companies that most often run the risk of being regarded as marginal, or too much of a challenge to detect. However, strengthening the link between sustainability projects and programs to the core business strategy of a company is probably more important than quantifying sustainability's contribution to corporate value drivers (Salzmann et al. 2005).

The very act of identifying and describing CSM's contribution to corporate value drivers captures the economic relevance of sustainability issues and can anchor this process to strategic development within companies. This (1) ensures that companies do not treat sustainability in such a tangential manner that they remain vulnerable to major shifts in what is, at the end of the day, a *socially driven* business environment; and (2) assures long-term sustainable growth in shareholder value.

As with all business cases, the case for sustainability has to be carefully built—for impact, companies need to rethink old business models and even venture toward becoming a pioneer ready to radically innovate. But, such innovations must reap bottom line benefits and be "value"-oriented:

> It does not work if the point is sustainable development, with no money-making. It is important to find win-win situations where the aim is to, at least, not lose money, and at the same time contribute to sustainable development.
>
> Senior business manager

Strategically Aligning behind the Business Logic

In implementing sustainability strategies, the challenge is relieving the tension between the complexity of sustainability issues, and the struggle to match and manage that complexity internally with frameworks, systems, tools, structures, processes, and the necessary human resources and values that allow smooth integration of sustainability into strategies.

Notwithstanding this challenge, from supply chain challenges to marketing techniques, sustainability can lead to new ways of working and reflecting on how companies do business; yet F&B companies are often anchored to "tried and true" business models, particularly in marketing and sales departments where short-term pressure is at its most intense. Business cases are currently interpreted very differently from company to company, from manager to manager, and roll-out is currently slow and halting, although constant. "Baby steps" rather than the "giant steps" required for a truly sustainable food system are the order of the day. No "leapfrogging" effect is forthcoming:

> The question for us is always, 'How fast and how far?' We are mainly taking small steps.
>
> Senior sustainability officer

While CSM in the industry may well pay back in the long run, accelerating efforts is not making business sense to many companies in the short- to medium-term. There are different business cases for different situations that tend to be very company-specific depending on a number of factors (type of product, cultural context, type of market, company history, level of public exposure, brand recognition benefits, and so on). But cases are often not fully exploited because managerial knowledge and awareness, tools, and guidelines needed to identify issues (and therefore risks) are lacking. The sector is lagging behind—little is going on to link sustainability risks across different functions within companies. Furthermore, incentive systems remain focused on short-term results, demotivating staff from tackling sustainability issues in a more hands-on way. Training and competence-building programs are not yet fully oriented toward reinforcing value systems needed to make CSM an integrated part of business.

The Center for CSM found that primary barriers to strategic rollout are thus internal factors completely within the company's control, indicating considerable potential for further exploitation of the existing business case. The F&B business environment is far too volatile for a "lackadaisical" approach to its multiple risks to continue for much longer. Top companies have started to move away from seeing sustainability from a purely cost-oriented perspective and are in a phase of exploring value-creating potential.

Conclusion

Although sustainability gives an opportunity to some (particularly branded) companies to establish rare, firm-specific, inimitable characteristics that can be the source of sustained competitive advantage so promoted by academics and sustainability heroes alike, too few companies are truly working on this as a genuine strategically integrated objective. The incremental approach we observed in companies before the financial crisis indicated that they do indeed have these issues on their agenda and are starting to manage them systematically. A highly positive indicator is the fact that existing efforts, while not increased, were at least not dropped off the radar screen during the worst of the crisis. One can only hope that these efforts and their correlative financial commitments will be renewed with stronger vigor than ever before once the global economy recovers.

As Plutarch once said, "Steady, continuous efforts followed by periodic but purposeful thrusts are irresistible, for this is how time captures the greatest powers on earth" (Plutarch c.46–120 CE).

Will the crisis provoke more of the required "purposeful thrusts" necessary? What is the verdict? The jury is still out on the answer to that question.

Notes

1. See http://www.imd.ch/research/centers/csm/index.cfm.
2. http://www.saiplatform.org.
3. http://www.sustainablefoodlab.org.

References

Billington, C., A. Ionescu-Somers, and M. Barnett Berg. 2009. Fill in the blanks on CSR. *CPO Agenda* (Summer, 2009). http://www.cpoagenda.com/previous-articles/summer-2009/features/fill-in-the-blanks-on-csr/ (accessed May 1, 2010).

de Fraiture, C., and D. Wichelns. 2010. Comprehensive assessment of water management in agriculture: Satisfying future water demands for agriculture. *Agricultural Water Management* 97 (4): 502–511.

Giles, M. 1993. After the feast. *Economist* 329, 7840 (December 4): 3–5.

Ionescu-Somers, A. 2006. Corporate customers and suppliers: How companies influence other companies on corporate sustainability. In *Inside the mind of the stakeholder: The hype behind stakeholder pressure*, ed. U. Steger. Hampshire, UK: Palgrave Macmillan.

Ionescu-Somers, A., and U. Steger. 2006. *Revenue flow and human rights: A paradox for Shell Nigeria.* IMD case study. Lausanne, Switzerland: IMD.

———. 2008. *Business logic for sustainability: A food and beverage industry perspective.* Hampshire, UK: Palgrave Macmillan.

Porter, M. E. 1985. *Competitive advantage: Creating and sustaining superior performance.* New York: Free Press.

Prinzhorn, J., and O. Salzmann. 2006. NGOs: Catalysts of corporate sustainability. In *Inside the mind of the stakeholder: The hype behind stakeholder pressure*, ed. U. Steger. Hampshire, UK: Palgrave Macmillan.

Reinhardt, F. L. 1999. Bringing the environment down to earth. *Harvard Business Review* (July-August): 149–157.

Salzmann, O., U. Steger, and A. Ionescu-Somers. 2005. *Quantifying economic effects of corporate sustainability initiatives: Activities and drivers.* Lausanne, CH: IMD Working Paper Series 2005-28.

Steger, U. 1998. *The strategic dimensions of environmental management.* London: Macmillan Press.

———. (Ed.). 2004. *The business of sustainability: Building industry cases for corporate sustainability.* Hampshire, UK: Palgrave Macmillan.

———. (Ed.). 2006. *Inside the mind of the stakeholder: The hype behind stakeholder pressure.* Hampshire, UK: Palgrave Macmillan.

Steger, U., A. Ionescu-Somers, and O. Salzmann. 2005. The business case for corporate sustainability: Literature review and research options. *European Management Journal* 23 (1): 27–36.

Steger, U., A. Ionescu-Somers, O.Salzmann, and S. Mansourian. 2009, *Sustainability partnerships: The managers handbook.* Hampshire, UK: Palgrave Macmillan.

CHAPTER 3

Sustainability as a Business Model

Wendy Stubbs

Introduction

Reports from the Intergovernmental Panel on Climate Change (IPCC) (2007), United Nations Environment Programme (UNEP) (2005), and the Worldwatch Institute (WWI) (2009) paint a bleak picture of the impact of human activity on the natural environment (see Table 3.1). While society and business rely on a healthy environment, the way humans use resources and the amounts used are irrevocably damaging the environment, such that humanity's very life support systems are under threat (McPhail 2008). Referencing the work of eminent British economist Sir Nicholas Stern (2006), Espinosa and colleagues conclude that "[t]he possibility that complete environmental collapse is now decades rather than millennia away makes it in everyone's interests to re-think fundamentally the structures and processes which have brought us so close to the brink" (Espinosa, Harnden, and Walker 2008, p. 645). McPhail (2008) concludes that business-as-usual is not working. Incrementally improving the environmental efficiency of existing businesses—producing the same amount of products but with fewer resources and less energy, waste, and emissions—is not enough. Ryan (2008) argues that the level of reduction required in greenhouse gas emissions points to the need for rapid systemic change in business systems because the traditional management paradigm is inherently limited in its ability to address ecological degradation effectively (Shrivastava 1995) and traditional business models exacerbate the problems (Stubbs and Cocklin 2008a).

Waddock (2007, p. 544) argues that while "interconnectedness and limitations of resources have dramatically changed the rules of the game," business thinking and practices are still rooted in the 1960s and 1970s. Kelly and White (2009) maintain that business leaders operate within a corporate

Table 3.1 Environmental impacts of human activity

Major changes to Earth's ecosystems (UNEP, 2005)	*Impacts of climate change (IPCC, 2007)*	*State of the World (The Worldwatch Institute, 2009)*
Half the world's wetlands were lost during the last century.	Global average surface temperature predicted to increase between 1.4 and 5.8 degrees Celsius from 1990 to 2100.	Unabated, current increasing trends in emissions can be expected to raise Earth's temperature by 4–6 degrees Celsius above today's levels, if not more, by the end of this century.
Logging and land use conversion have reduced forest cover by at least 20%, and possibly as much as 50%.	Emissions are projected to increase by up to 90% by 2030, and if they continue to rise, surface temperature will rise by up to 5 degrees by 2050.	A significant number of "tipping points"—thresholds beyond which it would become difficult-to-impossible to reverse changes in the climate system—could be approached if the planet warms more than 3 degrees. "Safe" levels of warming lie at 2 degrees Celsius or below.
Nearly 70% of the world's major marine fish stocks are either overfished or being fished at the biological limit.	Almost 1/3 of the world's species are predicted to face extinction if greenhouse gases continue to rise.	Once greenhouse gas concentrations are stabilized, global mean temperature will continue to rise due to momentum in the climate system for several decades.
Over the past half century, soil degradation has affected 2/3 of the world's agricultural land.	Water shortages will be exacerbated in many water-scarce areas of the world.	Observed rapid loss of Arctic summer ice exceeds projections in nearly all the latest IPCC models.
Each year, an estimated 27,000 species disappear from the planet.	20% of the world's population will face a great risk of drought.	The Gangotri Glacier in the Himalayas, which provides up to 70% of the water in the Ganges River, is retreating 35 meters yearly.
Dams and engineering works have fragmented 60% of the world's large river systems and severely impeded water flow.	Sea-level rise and storm surges will endanger small islands and low-lying coastal areas leading to increasing displacement of people.	Approximately 2/3 of the energy fed into the world's power plants is wasted—released into the environment as heat.
Human activities are significantly altering the basic chemical cycles upon which all ecosystems depend.		

design largely inherited from the nineteenth century when resources were plentiful, population was small, and humanity was not confronted with the ecological limits identified by the IPCC, UNEP, and WWI reports. The current business models are increasingly outmoded in the twenty-first century (Kelly and White 2009).

Some progress is being made, however. Birkin and colleagues' research provides some examples of organizations using business models that have changed significantly from those of a decade ago (Birkin, Polesie, and Lewis, 2009b). Companies are implementing environmental management systems, many in accordance with the ISO140001 standard, and applying the Global Compact and Global Reporting Initiative (GRI), none of which existed before 1996. Waddock (2008) reinforces this view by describing emerging institutions that are embedding an array of social, sustainability, and stakeholder issues into companies' business models, drawing on an "institutional infrastructure" encompassing the Global Compact, Millennium Development Goals, the Greenhouse Gas Protocol, ISO140001, and the GRI. Over the long term, Waddock (2008, p. 87) argues, these initiatives and trends will influence what companies need to do to "sustain their legitimacy and be accepted social actors."

Despite this progress, Birkin and colleagues (2009b, p. 278) conclude that there are still "unavoidable structural inhibitions in contemporary business models" that prevent companies from becoming sustainable, and fundamental change to traditional business models is needed to respond to societal, natural, and business needs of sustainable development (Birkin, Cashman, Koh, and Liu 2009a; Stubbs and Cocklin 2008a). In addition, the majority of effort of organizations to address sustainability issues (such as reducing emissions and waste, recycling, reducing energy usage, using alternative and more benign sources of energy, and developing greener consumer products) is considered an "add-on" to what remain essentially unsustainable business operations (Markevich 2009).

The greatest challenge facing business leaders today may well be the need to develop new business models that "accentuate ethical leadership, employee well-being, sustainability, and social responsibility without sacrificing profitability, revenue growth, and other indicators of financial performance" (Fry and Slocum Jr. 2008, p. 86). As such, this challenge is an important area for research and one that management researchers are urged to take up (Marcus and Fremeth 2009).

In response to a growing recognition that business must move from "sustainability as an add-on" to "sustainability as a business model," this chapter presents the "sustainability business model" (SBM)—a model where sustainability concepts shape the mission and driving force of the firm and its decision making (Wicks 1996). The SBM is an abstract model of a sustainable organization and was derived from in-depth case studies of two organizations that are considered leaders in implementing alternative sustainability-centric business models. The SBM can be used as a blueprint, or a design guideline, for new or existing organizations (Doty and Glick 1994). The chapter first reviews the literature that engages the debate about the need to move from sustainability as an add-on to sustainability as a business model. The chapter then explains the methods used to develop an "ideal type" (Blaikie 1993; Weber, Shils, and Finch 1949) SBM and describes the two case studies from which the SBM was derived. The chapter presents the characteristics and attributes of the SBM and, finally, it considers future directions and alternative paths to achieving sustainability as a business model.

The Challenges to Corporate Sustainability

Defining what corporate sustainability is, and therefore what "sustainability as a business model" means, is problematic. Definitions of sustainability are rife in the literature (Robinson 2004), with key constructs and alternative definitions proliferating during the past decade (Montiel 2008). However, there is consensus that corporate sustainability is concerned with balancing, or integrating, three spheres: economic responsibilities with social and environmental ones (Elkington 1997; Marcus and Fremeth 2009; Montiel 2008).

A key challenge for corporate sustainability is the perception that sustainability and the way we do business are fundamentally incompatible (Shrivastava 1995) in areas such as the inability to deal with externalities ("unpriced benefits or costs" such as pollution, the impact of global warming and climate change, erosion and depletion of natural resources) (Daly and Cobb 1994; Pearce 1978); treating environmental resources as "free"—ecosystem services and natural resources are not valued (Costanza, Daly, and Bartholomew1991; Pearce, Markandya, and Barbier 1989); and discounting the future which encourages short-term consumption of natural resources (Daly and Cobb, 1994; Pearce, Markandya, and Barbier 1989). Korhonen and colleagues (2004) suggest that the beliefs and values underlying sustainability, such as diversity, cooperation, community, connectedness, and locality run counter to "business-as-usual" (mass production, unlimited economic growth, competition, and globalization). According to Sharma and Starik (2004), the prevailing business model causes environmental and social degradation by encouraging organizations to externalize the social and environmental costs of their activities and internalize the economic benefits. Further to this, governments are not particularly effective at mitigating these externalized social and environmental costs of business activities.

Radical change is required (Espinosa, Harnden, and Walker 2008) as the traditional management paradigm was developed for the industrial society and is inherently limited in its ability to address ecological degradation effectively (Shrivastava 1995).

During the industrial revolution of the 1800s and early 1900s, population (900 million in 1800 and 1.6 billion in 1900 compared to 6.7 billion in 2008), resource consumption, waste, and greenhouse gas emissions were significantly less than they are today. As a result, the impact of human activity on the environment was relatively modest. The impact (I) of human activity on the environment was expressed as an equation by two environmental scientists, John Holden and Paul Ehrlich (1974): I = PAT. Environmental degradation is a direct result of the size of the world's population growth (P); what we consume—our level of affluence (A); and the amount of damage done through environmentally disruptive technology (T) such as fossil-fuel-driven technology and processes, and chemical pesticides.

Humanity's impact on the environment will increase as population, affluence, and the use of fossil-fuel technology increases. Holden and Ehrlich (1974, p. 291) concluded that ecological disaster will be difficult enough to avoid even if population is limited: "[I]f population growth proceeds unabated, the gains of improved technology and stabilized per capita consumption will be erased, and averting disaster will be impossible." Global population is expected to exceed 9 billion by 2050. Unlimited global economic growth is leading to a situation of "overshoot" where human demand continually exceeds the earth's regenerative capacity (Wackernagel et al. 2002). Increasing population and consumption will only exacerbate this situation. As population grows from 6.7 billion to over 9 billion, global economic growth is anticipated to multiply by three times over the next fifty years with a corresponding increase in resource consumption (Hamilton 2003). Hamilton (2003) estimates that to keep the impact of economic activity on earth at its current "destructive" level, technological advances will have to reduce resource use per unit of output to one quarter of the current levels. Doing so would pose enormous challenges for corporate sustainability, when the current business model is embedded in, and relies on, unlimited economic growth and increasing consumption. While new business models need to address these fundamental issues, there is some pessimism that such models will be developed and used. The UK Government's Sustainable Development Commission stated that "under current market conditions, it is almost impossible for businesses to engage seriously with any discussion about reducing levels of consumption" (cited in Birkin, Polesie, and Lewis 2009b, p. 278).

Birkin and colleagues (2009b, p. 288) report that there have been several attempts to define a business model that prioritizes sustainability, such as the

Nordic Partnership (2002), but they conclude that the initiatives do not involve a radical change to the prevailing economic-focused business model, and the initiatives "are not integrated by a comprehensive new understanding that could be identified as a new business model." Kelly and White (2009, pp. 25–26) point to successful companies that are implementing "alternative company designs" and propose six principles as a foundation for designing new corporate architectures that embrace sustainability: (1) the purpose of a corporation is to harness private interests to serve the public interest; (2) fair returns will accrue to shareholders but not at the expense of other stakeholders; (3) operations will be conducted sustainably; (4) wealth will be distributed equitably among those that contribute to its creation; (5) governance will be participatory, transparent, ethical, and accountable; and, (6) the right of people to govern themselves and other human rights will not be infringed.

This chapter builds upon these early studies and conceptual approaches by reporting a research project conducted in 2002–2005 seeking to understand what a sustainability-centric business model might look like.

Method

The research uses the case study method (Yin 2003). Case studies are an appropriate method for theory-building when little is known about the phenomenon under investigation (Eisenhardt 1989). The Websites, annual reports, and other public documents of organizations recognized for their sustainability initiatives were reviewed to select organizations that are implementing sustainability-centric business models, rather than treating sustainability as an "add-on" to their businesses. Initial sources of potential organizations included the Australian Corporate Responsibility Index, the Dow Jones Sustainability Index, the FTSE4Good index, and discussions with officers of Monash Sustainability Enterprises, an organization that undertakes research into socially responsible investment. Two organizations agreed to participate in the research study. Interface, a global carpet manufacturer, is considered to be a leader in restructuring its business model around environmental sustainability (Doppelt 2003; Elkington 2001). Bendigo Bank, Australia's sixth-largest bank, is implementing a community-development business model that focuses on building "successful" communities. Bendigo Bank was recognized as the Most Sustainable Company in Australia in 2001 and 2002 by Ethical Investor Magazine and received a merit award in 2003 for Outstanding Achievement in Social Development. Both companies are of theoretical interest because they are atypical (Lawrence 2002).

Data were collected in the period 2002–2005 from in-depth interviews with staff engaged in sustainability initiatives and from other secondary

Table 3.2 Sources of data

Data source	*Type of data*	*Bendigo Bank*	*Interface*
Semistructured interviews with employees, recorded and transcribed	Primary	11 interviews with the CEO and staff from the following departments: retail, solutions, information technology, strategy, community and alliance banking, and finance	10 interviews with staff from operations, manufacturing, sales and marketing, finance and IT, services, sustainability management, an environmental analyst, a VP and the founder/chairman
Semistructured interviews with stakeholders, recorded and transcribed	Primary	3 interviews with chairmen of community banks	No
Annual reports and CSR/sustainability reports	Secondary	2002–2004, 2009	1999–2003, 2008
Quarterly earnings announcements	Secondary	2004–2005, 2009	2003–2005, 2009
Internal company documents and personal communications	Secondary	Profile of organization; strategy; community bank principles; community bank policy and procedures; e-mails	E-mails
Organizations' Websites	Secondary	www.bendigobank.com.au	www.interfaceglobal.com
Newspaper articles, journal articles, and books	Secondary	Press articles 1998–2005, Community bank prospectuses, Research reports (Byrne, et al., 2003; Maine, 2000), Masters thesis (Moore, 2002)	Press articles 2002–2005, Books (Anderson, 1998; Doppelt, 2003; Dunphy, et al., 2003; Elkington, 2001; Griffiths, 2000; Rowledge, et al., 1999)
Annual general meeting and analyst briefing	Primary	2004	No

sources (see Table 3.2 for a summary of data sources). Twenty-four interviews were conducted; fourteen for Bendigo Bank (eleven Bendigo Bank staff members and chairmen from three community banks) and ten interviews were conducted with Interface staff (seven Australian staff and three staff from the head office in the United States). The objective in interviewing several employees within each organization was to acquire a breadth and depth of views and situations across each company. Grounded theory techniques (Glaser and Strauss 1967; Strauss and Corbin 1998) were used to code and analyze the data and identify the key themes emerging from the data. Theory is generated through creating an "ideal type" from the themes (Blaikie 1993).

Ideal types describe abstract models of the meanings used by social actors engaged in courses of action in certain situations (in this case, in situations where sustainability-centric business models are being implemented). An ideal type is a representation of the "idea" to the extent that "it has really taken certain traits . . . from the empirical reality of our culture and brought them together into a unified ideal-construct" (Weber, Shils, and Finch 1949, p. 91). The SBM is an ideal type that describes an abstract model of sustainable organizations, drawing on the meanings used by members of organizations who are employed in sustainability activities. According to Doty and Glick (1994), ideal types represent organizational forms that might exist rather than actually exist, but they can be used as design guidelines for new or existing organizations.

One limitation of using a small number of case studies is that doing so will not allow for generalization to a population (Blaikie 2000). However, making such generalizations was not the intention of the research study. The aim was to generate a theoretical understanding that can be used to develop an SBM. The theory that emerges during this process can be tested in other contexts to establish its range of application (Blaikie 2000).

Interface Case Study

Ray Anderson founded Interface in 1973 in Atlanta, Georgia, to produce and market modular soft-surfaced floor coverings. It is now the world's leading supplier of modular carpet (carpet tiles). It sells its products in over a hundred countries, primarily in the business market. Interface is a publicly listed company, on the NASDAQ exchange, and generates over US$1 billion of revenue annually.

To achieve its sustainability vision to be the first company that, by its deeds, shows the entire industrial world what sustainability is in all its dimensions, Interface developed a model of the "prototypical" company of the twenty-first century. This model embodies Interface's view of a sustainable enterprise: strongly service-oriented, resource-efficient, wasting nothing, solar-driven, cyclical rather than linear, and strongly connected to its constituencies (community, customers, and suppliers). The prototypical company seeks to

Table 3.3 Interface's environmental and social performance as of the end of 2009

Environmental metrics (EcoMetrics)	*Result*
Cumulative savings from global waste activities since 1995	US$433 million
Decrease in total energy consumption required to manufacture carpet since 1996	43%
Percentage of total energy consumption from renewable sources	30%
Reduction in direct GHG emissions	44%
Reduction in GHG emissions including offsets	71%
Reduction in water intake per square metre of carpet (modular) since 1996	77%
Reduction in water intake per square metre of carpet (broadloom) since 1996	47%
Amount of material diverted from landfill since 1995	200 million pounds
Decrease in manufacturing waste sent to landfill since 1996	80%
Percentage of recycled or bio-based content in products worldwide	36%
Safety—reduction in frequency of injuries since 1999	63%
Social metrics (SocioMetrics)	*Result*
Number of employee/family social events worldwide in 2009 (2001) The range of employee/family social events worldwide from 2001–2009 was 52 (2007) to 243 (2003).	69 (133)
Average hours of training per employee in 2009 (2001) The range of training hours per employee from 2001–2009 was 6.7 (2001) to18.7 (2008).	13 (6.7)
Contributions to charitable organizations in 2009 (2001) The range of contributions per year from 2001–2009 was $252,900 (2002) to $835,064 (2007).	US$331,000 ($423,650)
Employee volunteer hours in community activities in 2009 (2001) The range of volunteer hours from 2001-2009 was 7,368 (2006) to 18,775 (2008).	9,057 (12,714)

http://www.interfaceglobal.com/Sustainability.aspx (accessed April 28, 2010) and personal communication
Abbreviation: greenhouse gas (GHG)

go beyond complying with regulations, taking nothing from earth's crust (lithosphere) that is not renewable and not harming earth's biosphere (all living beings together with their environment) (Anderson 1998).

To achieve this business model, Interface is implementing "seven fronts" of sustainability (Stubbs and Cocklin 2008b):

1. Eliminate waste: Eliminate all forms of waste in every area of business.
2. Emit benign emissions: Eliminate toxic substances from products, vehicles, and facilities.

3. Use renewable energy: Operate facilities with renewable energy sources—solar, wind, landfill gas, biomass, geothermal, tidal, and low impact / small-scale hydroelectric or non-petroleum-based hydrogen.
4. Close the loop: Redesign processes and products to close the technical loop using recovered and bio-based materials.
5. Provide resource-efficient transportation: Transport people and products efficiently to eliminate waste and emissions.
6. Sensitize stakeholders: Create a culture that uses sustainability principles to improve the lives and livelihoods of all stakeholders—employees, partners, suppliers, customers, investors, and communities; and,
7. Redesign commerce: Create a new business model that demonstrates and supports the value of sustainability-based commerce that delivers service and value instead of material (product).

Interface has made significant progress since it first initiated its sustainability programs in 1994 (see Table 3.3), but believes it is only about halfway to achieving its vision. Table 3.3 shows substantial progress on all of the *environmental metrics* from the mid-1990s to 2009. Progress on the social metrics has been patchier. Although not clear from Table 3.3, the full set of 2001–2009 data suggest some upward trend for training and charitable contributions, a downward trend for social events, and no trend for volunteer hours. The *social metrics* are influenced by economic and business conditions, as is demonstrated by the decline in 2009 from peaks in earlier years.

Bendigo Bank Case Study

Bendigo Bank, which converted from a building society to a bank in 1995, is based in the central Victorian city of Bendigo, 150 kilometers northwest of Melbourne, Australia. It is a publicly listed company with a market capitalization that ranks it among the top hundred Australian companies on the Australian Stock Exchange. A retailer of banking and wealth management services to households and small-to-medium businesses, it holds assets under management of more than AU$47 billion and has over AU$28 billion in retail deposits.

During the 1990s when the "big four" Australian banks closed one-third of their bank branches, Bendigo Bank was undergoing a strategic review in response to an increasingly concentrated and competitive market. It sought to differentiate itself and to find a unique value proposition for customers and communities. It concluded that for it to thrive in its market—primarily regional banking—it needed to contribute to building more sustainable communities. It developed a wealth-sharing business model, the community engagement model (CEM). Community Bank is the first business based on the CEM. It is a branch-banking model that involves local people in meeting their own banking needs.

A community bank branch is owned by the local community, not by Bendigo Bank, and its establishment is driven by the local community—Bendigo Bank does not target or approach communities to open a community bank branch. Once the community has shown sufficient interest, a community steering committee is formed and a public company with limited shares is established. Shares are issued to raise funds from the local population to cover the setup costs and initial running costs of a community bank branch. These costs are about AU$600,000–$700,000. Each branch operates as a franchise of Bendigo Bank, using the name, logo, and system of operations of Bendigo Bank. The community company secures the branch premises, purchases fittings and systems, and covers the branch running costs such as wages, power, and telecommunications. Bendigo Bank provides the banking license, the bank brand, training of staff, a core range of products and services, systems, marketing support, and administrative support. The community bank branch and Bendigo Bank share the revenue from the products and services sold through the community bank. The margin varies across the products and services depending on the amount of work done by each party. Generally, the community bank branch receives about 50 percent of the revenue from consumer products but less on business products (Stubbs & Cocklin 2007).

Of a community bank's profit, 20 percent can be distributed as dividends to its shareholders, once the community bank has accumulated a net profit (when previous years' losses are cleared). Only local community members can be shareholders in the community bank. Of the profit, 80 percent is set aside to fund further community development projects. This

Table 3.4 Bendigo Bank results from its community banking model (all amounts in AU$)

Metric	*Result*
Number of community bank branches (as in March 2010)	256
Total amount of revenues received by community banks (as of the end of 2007)	$340 million
Amount of revenue paid to community banks each year (as of 2009)	$100 million
Amount paid to community bank shareholders in dividends	$11 million
Amount contributed to community projects	$30 million
Number of community bank customers	410,000
Number of local jobs created	1,500
Number of community bank shareholders	61,000
Amount community banks directly inject into their communities each year (through salaries and other local expenditure)	$43 million

Bendigo and Adelaide Bank Annual Report 2009 and personal communication eighty/twenty split is built into the community bank franchise agreement. One employee explained that the reason for this split was "to ensure there is a balance between being cooperatively spirited and commercially based." The community bank board of directors decides how to distribute the profits and which community projects to fund (Stubbs and Cocklin 2007). The outcomes from the CEM since Bendigo Bank opened its first community bank branch in 1998 are summarized in Table 3.4.

Findings: The Sustainability Business Model

The key themes arising from the analysis of the case studies that inform the SBM are summarized below (Stubbs and Cocklin 2008a). The themes highlight the sustainability competencies that make up the SBM. "Competencies allow an organization to tie together complementary and cospecialized capabilities" such as newspaper, plastic, and paper recycling, consumer education, advanced recycling, and offering environmental products and services (Marcus and Fremeth 2009, p. 22).

The Purpose of Business and the Role of Profits

The SBM draws on economic, environmental, and social aspects of sustainability to define an organization's purpose. Both Interface and Bendigo Bank defined the purpose of their businesses in terms wider than purely financial ones (profitability and shareholder returns). Interface's purpose emphasized environmental aspects (cherishing nature and restoring the environment) and social aspects of sustainability (maximizing all stakeholders' satisfaction), while Bendigo Bank focused on the social aspects (improving the prospects of customers and communities). Both acknowledged that profits are an outcome, and a facilitator, of environmentally and socially sustainable activities. In pursuing this approach, they face challenges in changing cultures and attitudes internally and externally, to get buy-in and proactive support from their stakeholders, such as staff, the board of directors, shareholders, business partners, customers, communities, and financial market analysts. They are both pursuing sustainability for ethical (it is the "right thing to do") and economic reasons—"it is the right thing to do and the smart thing to do" (Interface). However, according to Interface, "short-termism" is a real challenge. One interviewee pointed out that it could be "in conflict with the long term sustainability focus." Bendigo Bank's experience is that sustainability initiatives can build long-term value for its shareholders and it is willing to "temper dividends" in the short term to continue to invest in its community engagement initiatives. Birkin and colleagues (2009b, p. 287) also concluded that for a new business model, the "good" or "right" thing to do may lie outside the finance-driven model "with the consequence that other bases for business decision-making need to be identified."

Wealth Sharing and Local Capital

A key driver of Bendigo Bank's community engagement strategy is to retain capital in communities through revenue sharing, local shareholder ownership, and reinvestment of profits. The key features of "retaining local capital" that are important conditions for an SBM include the motivation to work for the "common good" for the benefit of multiple stakeholders not just shareholders; a willingness to work cooperatively with stakeholders to achieve social and environmental outcomes; and a willingness to temper short-term financial outcomes so that social and environmental outcomes can be achieved. The first two themes are in concert with Kelly and White's (2009) first (purpose), second (capital), and fourth (wealth) principles of corporate design, discussed earlier in this chapter.

Sustainability Performance

Interface uses a Triple Bottom Line (TBL) approach in reporting its performance. The TBL (Elkington 1997) encompasses the economic, environmental, and social performance of an organization. Bendigo Bank publishes information on its social, environmental, and financial performance but does not use TBL reporting. The use, or not, of TBL reporting is not particularly significant in itself. For example, companies may report their progress on recycling, levels of emissions, and community involvement initiatives but may not be changing their underlying business practices that cause environmental and social degradation. Likewise, companies may not use a TBL reporting format but may be making significant progress toward sustainability, as demonstrated by Bendigo Bank. This observation suggests that TBL reporting by itself is not a good indicator of sustainability and it is not a sufficient and necessary condition for companies to achieve sustainability. Reporting progress on sustainability influences stakeholders' perceptions and is therefore an important tactic, but on its own it does not appear to be a necessary driver of sustainability.

Stakeholders

An SBM considers the needs of all stakeholders (human and natural), rather than giving priority only to shareholders' expectations. Stakeholder engagement is a major strength of Bendigo Bank; it is central to its CEM. Bendigo Bank's model is based on building trust through cooperating and collaborating with its communities and customers. It seeks out partnerships with other organizations to find ways to meet communities' needs (for example, it has partnered to launch eight community telecommunications companies). Interface is at an earlier stage of stakeholder engagement (for example, it addressed community needs only when resources were slack) but it does acknowledge that nature is a stakeholder, recognizing that the natural environment is a vital component of the business environment (Starik 1995). Interface and Bendigo Bank believe that many of their shareholders invest in them because of their social and environmental sustainability agendas; their "bigger than itself" strategies (Bendigo Bank). On the basis of this standpoint, a more typical scenario for an SBM is that shareholders and the financial investment community recognize that sustainability initiatives build long-term value for all stakeholders. This perspective may mean accepting a lower return on investment (dividends) in the short term (which Bendigo Bank shareholders have done) so that organizations can direct profits to programs to support social and environmental initiatives. An SBM requires that the values of shareholders, stakeholders, and organizations are aligned around sustainability outcomes, which may only occur when sustainability is institutionalized in society (Jennings and Zandbergen 1995).

Structural and Cultural Competencies

Both organizations highlighted the importance of leaders or champions in driving the changes in culture (norms, values, and attitudes) and structure (organization structure, policies, and processes) required to create a sustainability-committed organization. In both case studies, it was the individual concerns of the CEO that initially drove the change process that shaped the organizational agenda around sustainability (Bansal 2003). Both CEOs successfully "sold" sustainability to stakeholders (the board, management, staff, shareholders, and customers) and the organizational values are now more aligned with the CEOs' sustainability values (in Bendigo Bank's case, the CEO stated that "community values" were always central to the bank's strategy). The interviewees from both organizations concurred that sustainability is now embedded in the culture and the role of the CEO in "converting" the organization to sustainability has diminished; there are now many champions within each organization. Both case studies suggest that until sustainability is institutionalized in organizations and within the mindsets of stakeholders, "visionary CEOs" will be required to push the sustainability agenda throughout organizations.

An SBM promotes environmental stewardship: renewable or human-made resources are used instead of nonrenewable resources / natural capital; technological innovation minimizes, and eventually eliminates, nonrecyclable waste and pollution; changes in consumption patterns are encouraged to reduce the "ecological footprint" of people and organizations; and environmental damage is repaired. Interface's experience suggests that structural changes, which require capital investment (for recycling plants, new manufacturing technology and processes, renewable energy facilities, and redesigned transportation systems), as well as behavioral changes (less consumption) are needed to achieve environmental sustainability. Bendigo Bank achieves social sustainability through more intangible factors, such as relationship, trust, and engagement building activities, and these may be more difficult to replicate than more tangible factors such as technology.

Internalizing Costs

Interface is redesigning its products and services to eliminate environmental damage and waste, and absorbs the costs of doing this. The SBM internalizes all costs associated with environmental and social degradation and damage. To build a business environment that fully supports the realization of the SBM, modifications to the taxation systems (such as through an emissions trading scheme or carbon tax) are required to encourage organizations to redesign their products and practices to eliminate negative environmental impacts. Such change would encourage investment in infrastructure to support recycling, "clean" energy, "clean" transportation, and closed-loop systems (to avoid environmental taxes). Interface stated that it is too small to lobby government for changes to the taxation system. Rather than direct lobbying, Interface's approach is to work with coalitions that have more power to lobby government such as the World Resources Institute and the World Business Council.

Figure 3.1 summarizes the characteristics of the SBM derived from the two case studies. Some characteristics can be partially achieved through internal capabilities but also require changes in the socioeconomic environment, such as eliminating waste and emissions, and implementing closed-loop systems. A long-term focus is required by both the organization and the socioeconomic environment. "Patient shareholders" is classified as both a structural (the ownership structure includes patient shareholders) and a cultural ("patient" is a cultural attribute of shareholders) characteristic.

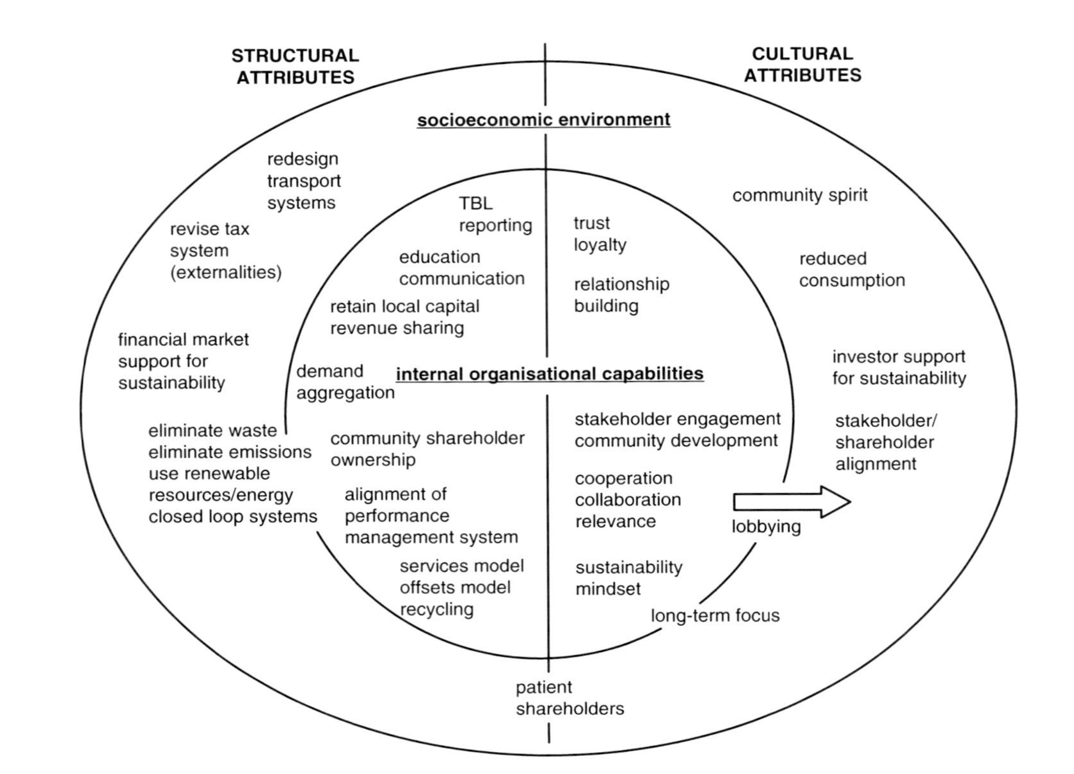

Figure 3.1 The characteristics of the Sustainability Business Model

Conclusion

While drawing theoretical conclusions from the particulars of only two case studies is inherently risky (Lawrence 2002), the analysis of the data highlighted a number of challenges to move from "sustainability as an add-on" to "sustainability as a business model." Organizations can make significant advances toward achieving sustainability through their own internal capabilities, but ultimately organizations can only be sustainable when the whole system that they are part of is sustainable. That is, changes are required at the socioeconomic system level, both structural (such as redesigning transportation systems and taxation systems) and cultural (such as changes to attitudes about consumption and economic growth). Interface in particular highlighted the fact that it cannot be fully sustainable until its whole supply chain is sustainable ("we are our supply chain"). An SBM requires stakeholders such as organizations, governments, industry bodies, supply chains, NGOs, and society to work together to address sustainability from a systems perspective. Kelly and White (2009) reinforce this approach by arguing that new corporate architectures require a combination of internal corporate initiatives, government mandates, and civil society pressures. What also became clear from the interviews was that an SBM would only emerge fully when sustainability is institutionalized in society; that is, when sustainability is addressed as a societal issue, rather than addressed separately by individual constituencies. Nevertheless, organizations can make substantial progress toward sustainability by adopting the SBM as a design guideline (Doty and Glick 1994). In doing so, the examples they set can contribute to societal constituencies working toward a sustainable world.

The SBM is one ideal type of "sustainability as a business model." There are other pathways to sustainable business models such as Dixon and Clifford's (2007) "ecopreneurship" model—a social and environmental entrepreneurship business model; microfinance models (Hassan, 2002)—providing access to credit for underprivileged groups that might not otherwise have the education or collateral to obtain a loan—pioneered by Grameen Bank in the 1970s; and cooperatives that are autonomous associations of people united voluntarily to meet their common economic, social, and cultural needs and aspirations through a jointly owned and democratically controlled enterprise, such as the Co-operative Group in the UK and Mondragon in Spain. The priority for Mondragon is to increase jobs and preserve community, and it retains 90 percent of its annual profit to reinvest in other cooperative initiatives. The Co-operative Group's top priority is to provide the best possible services to its members and to invest in the communities where they live.

Further research is required to verify and expand the SBM as well as investigate ideal types of sustainable organizations within different contexts, such as developing countries (Boutilier 2005, Sharma and Starik 2004).

References

Anderson, R. C. 1998. *Mid-course correction*. Atlanta, GA: Perengrinzilla Press.

Bansal, P. 2003. From issues to actions: The importance of individual concerns and organizational values in responding to natural environmental issues. *Organization Science* 14(5): 510–527.

Bendigo and Adelaide Bank. 2009. *Building Customer Connections*. Full Annual Report 2009. Bendigo, VIC.

Birkin, F., A. Cashman, S. Koh, and Z. Liu. 2009a. New sustainable business models in China. *Business Strategy and the Environment* 18(1): 64–77.

Birkin, F., T. Polesie, and L. Lewis. 2009b. A new business model for sustainable development: An exploratory study using the theory of constraints in Nordic organizations. *Business Strategy and the Environment* 18(5): 277–290.

Blaikie, N. W. H. 1993. *Approaches to social enquiry.* Cambridge, MA: Polity Press Blackwell.

———. 2000. *Designing social research: The logic of anticipation*. Malden, MA: Polity Press.

Boutilier, R. 2005. Views of sustainable development: A typology of stakeholders' conflicting perspectives. In *New horizons in research on sustainable organisations: Emerging ideas, approaches and tools for practitioners and researchers*, ed. M. Starik, S. Sharma, C. Egri, and R. Bunch, 19–37. Sheffield, UK: Greenleaf Publishing.

Byrne, G., J. Francis, J. McGovern, and I. Pinge. 2003. *A socio-economic evaluation of community banking. Bendigo Bank and the Centre for Sustainable Regional Communities.* Bendigo: La Trobe University.

Costanza, R., H. E. Daly, and J. A. Bartholomew. 1991. Goals, agenda, and policy recommendations for ecological economics. In *Ecological economics: The science and management of sustainability*, ed. R. Costanza, 1–20. New York: Columbia University Press.

Daly, H. E., and J. B. Cobb. 1994. *For the common good: redirecting the economy toward community, the environment, and a sustainable future.* Boston, MA: Beacon Press.

Dixon, S. E. A., and A. Clifford. 2007. Ecopreneurship—a new approach to managing the triple bottom line. *Journal of Organizational Change Management* 20(3): 326–345.

Doppelt, B. 2003. *Leading change toward sustainability: A change-management guide for business, government and civil society*. Sheffield, UK: Greenleaf Publishing.

Doty, D. H., and W. H. Glick. 1994. Typologies as a unique form of theory building: Toward improved understanding and modeling. *Academy of Management Review* 19(2): 230–251.

Dunphy, D. C., A. Griffiths, and S. Benn. 2003. *Organizational change for corporate sustainability: A guide for leaders and change agents of the future.* London: Routledge.

Eisenhardt, K. M. 1989. Building theories from case study research. *Academy of Management Review* 14(October): 532.

Elkington, J. 1997. *Cannibals with forks: The triple bottom line of 21st Century business.* Oxford: Capstone.

———. 2001. *The Chrysalis Economy: How citizen CEOs and corporations can fuse values and value creation.* Oxford: Capstone.

Espinosa, A., R. Harnden, and J. Walker. 2008. A complexity approach to sustainability—Stafford Beer revisited. *European Journal of Operational Research* 187(2): 636–651.

Fry, L. W., and J. W. Slocum Jr. 2008. Maximizing the Triple Bottom Line through spiritual leadership. *Organizational Dynamics* 37 (1): 86.

Glaser, B. G., and A. L. Strauss. 1967. *The discovery of grounded theory: Strategies for qualitative research.* New York: Aldine.

Griffiths, A. 2000. New organisational architectures: Creating and retrofitting for sustainability. In *Sustainability: The corporate challenge of the 21st century,* ed. D. C. Dunphy, J. Benveniste, A. Griffiths, and P. Sutton, 219–235. St. Leonards, NSW: Allen & Unwin.

Hamilton, C. 2003. *Growth fetish.* Crows Nest, NSW: Allen & Unwin.

Hassan, M. K. 2002. The microfinance revolution and the Grameen Bank experience in Bangladesh. *Financial Markets, Institutions & Instruments* 11(3): 205–265.

Holden, J. P., and P. R. Ehrlich. 1974. Human population and the global environment. *American Scientist* 62: 282–292.

Intergovernmental Panel on Climate Change (IPCC) 2007. Climate change 2007: Synthesis report. *Working Group IV Report.* Geneva, Switzerland: IPCC.

Jennings, P. D., and P. A. Zandbergen. 1995. Ecologically sustainable organizations: An institutional approach. *Academy of Management Review* 20(4): 1015–1052.

Kelly, M., and A. L. White. 2009. From corporate responsibility to corporate design: Rethinking the purpose of the corporation. *Journal of Corporate Citizenship,* no. 33: 23–27.

Korhonen, J., F. von Malmborg, P. A. Strachan, and J. R. Ehrenfeld. 2004. Management and policy aspects of industrial ecology: An emerging research agenda. *Business Strategy and the Environment* 13(5): 289–305.

Lawrence, A. T. 2002. The drivers of stakeholder engagement: Reflections on the case of Royal Dutch/Shell. *Journal of Corporate Citizenship* 6: 71–85.

Maine, M. 2000. *Banks for people: Community banking—prospects and limits,* Perth, WA: Institute for Science and Technology Policy Murdoch University, ISTP occasional paper, 1/2000.

Marcus, A. A., and A. R. Fremeth. 2009. Green management matters regardless. *Academy of Management Perspectives* 23(3): 17–26.

Markevich, A. 2009. The evolution of sustainability. *MIT Sloan Management Review* 51(1): 13–14.

McPhail, I. 2008. *Victoria state of the environment report 2008.* Melbourne: Commissioner for Environmental Sustainability.

Montiel, I. 2008. Corporate social responsibility and corporate sustainability: Separate pasts, common futures. *Organization & Environment* 21(3): 245–269.

Moore, T. 2002. Community Bank branch investment: Motivations and consequences. *Department of Economics.* Melbourne: Monash University.

Nordic Partnership, The. 2002. *Business models for sustainability: Experiences and opportunities from the Nordic Partnership.* http://www.drabaekconsult.dk/LinkedDocuments/Business_model.pdf (accessed November 16, 2009).

Pearce, D., A. Markandya, and E. B. Barbier. 1989. *Blueprint for a green economy.* London: Earthscan Publications Limited.

Pearce, D. W. 1978. *The valuation of social cost.* Boston, MA: Allen & Unwin.

Robinson, J. 2004. Squaring the circle? Some thoughts on the idea of sustainable development. *Ecological Economics* 48(4): 369–384.

Rowledge, L. R., R. S. Barton, and K. S. Brady. 1999. *Mapping the journey: Case studies in strategy and action toward sustainable development.* Sheffield, UK: Greenleaf Publishing.

Ryan, C. 2008. Climate change and ecodesign: Part I: The focus shifts to ystems. *Journal of Industrial Ecology* 12(2): 140–143.

Sharma, S., and M. Starik. 2004. Stakeholders, the environment, and society: Multiple perspectives, emerging consensus. In *Stakeholders, the environment, and society*, ed. S. Sharma and M. Starik: 1–22. Northampton, MA: Edward Elgar.

Shrivastava, P. 1995. Ecocentric management for a risk society. *Academy of Management Review* 20(1): 118–137.

Starik, M. 1995. Should trees have managerial standing? Toward stakeholder status for non-human nature. *Journal of Business Ethics* 14(3): 207–217.

Stern, N. 2006. *Stern Review: The economics of climate change.* Cambridge: Cambridge University Press.

Strauss, A. L., and J. M. Corbin. 1998. *Basics of qualitative research: Techniques and procedures for developing grounded theory.* Thousand Oaks, CA: Sage Publications.

Stubbs, W., and C. Cocklin. 2007. Cooperative, community-spirited, and commercial: Social sustainability at Bendigo Bank. *Corporate Social Responsibility & Environmental Management* 14: 251–262.

———. 2008a. Conceptualizing a "sustainability business model." *Organization & Environment* 21(2): 103–127.

———. 2008b. An ecological modernist interpretation of sustainability: The case of Interface Inc. *Business Strategy and the Environment* 17(8): 512–523.

United Nations Environment Programme (UNEP). 2005. *One planet many people: Atlas of our changing environment.* Nairobi, Kenya: Division of Early Warning and Assessment (DEWA), United Nations Environment Programme (UNEP).

Wackernagel, M., N. B. Schulz, D. Deumling, A. C. Linares, M. Jenkins, V. Kapos, C. Monfreda, J. Loh, N. Myers, R. Norgaard, and J. Randers. 2002. Tracking the ecological overshoot of the human economy. *Proceedings of the National Academy of Sciences* 99(14): 9266–9271.

Waddock, S. 2007. Leadership integrity in a fractured knowledge world. *Academy of Management Learning & Education* 6(4): 543–557.

Waddock, S. 2008. Building a new institutional infrastructure for corporate responsibility. *Academy of Management Perspectives* 22(3): 87–108.

Weber, M., E. Shils, and H. A. Finch. 1949. *The methodology of the social sciences.* New York: Free Press.

Wicks, A. C. 1996. Overcoming the separation thesis: The need for a reconsideration of business and society research. *Business and Society* 35(1): 89–118.

Worldwatch Institute, The. 2009. *State of the world: Into a warming world.* Washington, D.C.: Worldwatch Institute.

Yin, R. K. 2003. *Case study research: Design and methods.* Thousand Oaks, CA: Sage Publications.

CHAPTER 4

Investing in Sustainable Entrepreneurship and Business Ventures in the Developing World: Key Lessons and Experiences of EcoEnterprises Fund

Tammy E. Newmark and Jacob Park

Introduction

Rapid growth of the global economy has accelerated the degradation of the Earth's most important asset: the environment. With the world's population increasing, the demand on natural resources to cover even basic human needs is intensifying, and finding ways to meet those needs without exhausting our finite resource base will be imperative. The good news is that there is an emerging realization that economic growth does not have to come at the environment's expense. Around the world, passionate entrepreneurs are introducing innovative products and ways of doing business that use natural resources sustainably and can help ensure their long-term viability.

Ryan and Jeremy Black, two brothers who introduced the newest energy juice to hit the United States health conscious markets, epitomize this breed of sustainable entrepreneurs. Through their company Sambazon, they have mainstreamed açai, a tiny Brazilian berry grown in the Amazon, which is rich in antioxidants. Sambazon is able to provide much needed income to Amazonian communities because the company encourages sustainable management and harvesting techniques for açai. For companies like Sambazon that care as much about the social and environmental impacts of their operations as they do about profits, start-up capital is hard to find. What turned Sambazon into a success was a timely, initial investment of $225,000 from EcoEnterprises Fund (EcoE), a unique venture capital fund formed by the Inter-American Development Bank's Multilateral Investment Fund and The Nature Conservancy (Newmark 2009).

Even with visionary entrepreneurs like Ryan and Jeremy Black, many green businesses never get off the ground because traditional sources of capital like banks tend to shy away from sectors that seem unfamiliar or too risky. This situation is unfortunate because poverty and environmental conservation critical to global sustainability remain two of the biggest challenges confronting the international community. Increasingly, there is a call for the business sector to respond strategically to these issues by capitalizing on market opportunities posed by the pressures of poverty and environmental degradation in emerging and developing economies. In doing so, businesses can contribute to a more sustainable world. Using the impact investment fund model of EcoE as a case study, this chapter explores the theory and practice of investing in sustainable entrepreneurship and business ventures in the developing world by examining two interrelated issues and questions.

First, what is the process, and how well do we, as academics and policy makers, understand the institutional mechanisms behind investing in sustainable entrepreneurship and business ventures? Moreover, how different or similar are these processes in developed countries versus the processes in developing and emerging economies? Although the academic scholarship review will address the context of developing and emerging economies in general, there will be a special focus on Latin America where EcoE makes its investments. Second, what important lessons do the experiences of EcoE hold for mainstreaming sustainable entrepreneurship and investments in the developing world?

Launched in 2000, EcoE is an impact investment fund that targets the emerging green business sector in Latin America and offers both risk capital and business advisory services to small- to medium-scale enterprises (SMEs) in ecotourism, sustainable agriculture, apiculture, sustainable forestry, and nontimber forest products. The term "impact investing" is used to describe EcoE's approach, although terminology and definitions surrounding the social and environmental dimensions of the investment field can be best described as rapidly changing. There are those who might see the work of EcoE as in the "social capital" arena, while others might prefer to use "environmental venture capital." The terms "sustainable entrepreneurship" and "sustainable investing" are used to describe in more generic sense the social and environmental links in the investment and entrepreneurship fields.

Sustainable Investment and Sustainable Entrepreneurship: Literature Review

Inquiry into the social and environmental impacts of businesses, particularly multinational corporations, is not new. Decades before the word "globalization" entered common usage, the late Raymond Vernon, Professor of International Affairs Emeritus at Harvard University, was researching and writing about the social impacts of private companies (Vernon 1977). What is new however is the proliferation of corporate social and environmental responsibility codes of conduct, standards, and guidelines directed at governing business behavior, particularly in the petroleum, mining, and other extractive industries (Frankental and House 2000). Even as recently as a decade ago, it might have been unusual to hear the words "corporations" and either "climate change" or "human rights" being uttered in the same sentence. Now, one can read articles that link corporations to these and similar issues in newspaper and magazines almost on a daily basis (Park 2004). According to a 2007 McKinsey & Co. global survey, 95 percent of CEOs believe that "society has greater expectations than it did five years ago that companies will assume public responsibilities" (Bielak, Bonini, and Oppenheim 2007, p. 1). Corporate environmental and social responsibility concerns have grown into a global phenomenon involving a wide range of stakeholders in businesses, governments, and civil society.

Sustainable Investing

This chapter examines and builds on two important components of the growing academic literature about corporate social and environmental responsibility: sustainable investing and sustainable entrepreneurship. In many ways, the emergence of sustainable investing highlights the complexities of contemporary global economic and environmental governance (MacLeod and Park, forthcoming): the blurring lines between the domestic and international (as investment capital increasingly attempts to link local corporate activities with global responsibilities); the expansion of investment actors (from a few dedicated mutual funds mainly in the United States to now several hundred if not thousands across the world); the variety of transnational issues motivating capital markets (from human rights–related concerns such as apartheid in South Africa to emerging environmental sustainability and related concerns in China); and the exercise of private authority in issue areas where public authority has traditionally been unwilling or unable to act (Haufler 2006).

In terms of business management and academic scholarship, the significance of sustainable investment is two-fold. First, there is increasing academic scholarship on a wide range of sustainable investment-related themes (Park, forthcoming), including the relationship between corporate social responsibility and socially responsible investment (Hill, Ainscough, Shank, and Manullang 2007); links between financial and social-environmental performance (Margolis and Walsh 2001); the ability of social ratings to measure corporate social responsibility (Chatterji, Levine, and Toffel 2007); and shareholder activism and corporate social performance (Parthiban, Bloom, and Hillman 2007) among others. A 2009 meta-analysis of 36 academic studies that examine the relationship between financial performance and environmental, social, and governance (ESG) factors reveals that 20 studies find a positive relationship (that is, financial and ESG factors are positively correlated), 8 studies show a neutral relationship, and 6 studies show a neutral/negative relationship (Mercer 2009). A 2006 survey of 183 large financial institutional investors conducted by Mercer Investment Consulting indicates that as many as 75 percent of the respondents (22 percent of whom are currently SRI investors) believe that social, environmental, and corporate governance factors can have a material impact on investment performance (Mercer 2006).

Second, individual and institutional investors are becoming increasingly transnational actors that not only provide investment capital in international equity markets, but also exert social and environmental sustainability influence. Investors are exerting their economic as well as social and environmental sustainability influence through the shares they own in particular companies, which gives them the right to bring forth shareholder resolutions, but influence the shares that they do *not* own by creating standards or screens for potential future equity purchases (MacLeod and Park, forthcoming). While it is difficult to determine the long-term impact and effectiveness of such sustainable investment activities as portfolio screening and shareholder advocacy/engagement practices, there is growing evidence that sustainable investing is becoming a mainstream investment concept. As noted in a 2009 social investing report commissioned by the Rockefeller Foundation: "Evidence suggests that many thousands of people and institutions around the globe believe our era needs a new type of investing. They are already experimenting with it, and many of them continue even in the midst of a financial and credit crisis. That's why the idea of using profit-seeking investment to generate social and environmental good is moving from a periphery of activist investors to the core of mainstream financial institutions" (Freireich and Fulton 2009, p. 5).

Sustainable Entrepreneurship

Definitions of social and environmental or sustainable entrepreneurship are varied. Mair and Ganly (2010, p. 104) define it as . . . "initiatives that proactively address social or environmental issues through delivery of a product or service that directly or indirectly catalyzes social change," while what sustainable entrepreneurs do as a core strategy can be regarded as "challenging or trying to change excessive consumption, environmentally unsustainable practices, and a culture of individual private gain over shared community or public benefit. . . . " Austin, Stevenson, and Wei-Skillern (2006, p. 1) define it as "entrepreneurial activity with an embedded social purpose," while Dean and McMullen (2007, p. 58) describe it as "the process of discovering, evaluating, and exploiting economic opportunities that are present in environmentally relevant market failures."

While the origin of the term "social entrepreneurship" can be traced to Bill Drayton, former business management consultant who started Ashoka in 1980 (Mair and Ganly 2010), there are various subcategories of the sustainable entrepreneurship term including "enviro-capitalists" (Anderson and Leal 1997); "ecopreneurship" (Schaltegger and Petersen 2001, Ivanko and Kivirist 2008), "environmental entrepreneurship" (Schaper 2002, Schaper 2005), and "corporate social entrepreneurship" (Austin, Leonard, Reficco, and Wei-Skillern 2006), among others.

Although there are conceptual disagreements on what constitutes social and environmental or sustainable entrepreneurship, there is consensus among scholars and business researchers that entrepreneurs can be defined by their strong desire to conceive new business opportunities and develop new products and/or services to the marketplace. As the academic entrepreneurship literature has reported, entrepreneurs seek to bring about change and new opportunities, and play the roles of change agents in their respective organizations and communities in introducing innovation and new ideas (Schumpeter 1934, Schumpeter 1950, and Drucker 1985). The archetypal entrepreneur is viewed as an individual who starts his or her own small business that eventually grows into a larger and more successful corporation, but entrepreneurs or "intrapreneurs" can also be found within existing large companies (SustainAbility 2008). Most importantly, entrepreneurship can be found in all societies and in all types of economic circumstances (Schaper 2002) with a strong focus on innovation (Larson 2000).

Another emerging consensus centers on the growing recognition of sustainable entrepreneurs as economic and business actors. While comparative data on sustainable enterprises are difficult to obtain because countries tend to define social and environmental entrepreneurship differently, the overall scope and economic impact is noteworthy, particularly in Western Europe. Italy created a legal form for "social cooperatives" in 1991, which has grown to include 7,000 such organizations employing 200,000 workers by 2001. The 2006 Global Entrepreneurship Monitor data for the UK reported that 3.3 percent of the UK population was involved in creating or running an early-stage social enterprise, while another 1.5 percent ran an established social enterprise (Mair and Ganly 2010).

Sustainable Investment and Entrepreneurship at the Base of the Pyramid

When the Grameen Bank and its founder Muhammad Yunus were jointly awarded the Nobel Peace Prize in 2006, the award highlighted what many people in the economic development community had known for a long time: sustainable investment and entrepreneurship has become an important vehicle of long-term economic development in emerging and developing economies. Books like David Borstein's *How To Change the World: Social Entrepreneurs and the Power of New Ideas* (2004) and C. K. Pralahad's *The Fortune at the Bottom of the Pyramid: Eradicating Poverty Through Profits* (2004) also posed important questions about what role the private sector and traditional business concerns like strategy and entrepreneurship should play at the base of the global economic pyramid where nearly 80 percent of the 6.8 billion people on this planet live. Most critically in terms of market development, SMEs can have a multiplier effect on the economy by accelerating employment; raising incomes; and helping build new products, services, and business models that fundamentally alter an industry (Yago, Roveda, and White 2007).

Problems of high financial inflexibility and low business scalability have traditionally been regarded as purely economic development issues, but they are now emerging as critically important environmental and social concerns as well. Forestry, fishing, farming, and other types of ecological resource and renewable-energy-based enterprises can provide a strong business foundation for delivering economic, environmental, and social benefits to the poor, while simultaneously maintaining if not improving the natural resource base of a local community. Under the right set of policy and market circumstances, it is possible for the poor to adhere to good resource and stewardship and business management practices while reducing poverty and building more resilient communities (World Resources Institute et al. 2008).

There is a critical need to develop such "sustainable" enterprises in emerging and developing economies where the greatest concentration of poverty exists. The core of the base of the pyramid concept is essentially true, that is, multinational corporations do need to pay attention to the market potential of those living and working at the base of the global economic pyramid. But, the challenge is to redirect that market demand toward serving the basic human development needs that will reduce poverty as opposed to just intensifying unsustainable consumption patterns that already exist in many industrialized and in some emerging economies. For instance, more than one-third of the world's population (2.4 billion people), the majority of whom live in emerging and developing countries, do not currently have access to clean and reliable sources of energy. To meet the basic cooking needs of these 2.4 billion people, it is estimated that no more than 1 percent of the current global commercial energy consumption would be required (UNESCAP 2005). However, can this 1 percent of current global commercial energy consumption be increased in a way that is economically, socially, and environmentally sustainable for those living and working at the base of the pyramid and for the international community?

Although SMEs represent the dominant form of business organization worldwide, accounting for more than 90–95 percent of business establishments, these companies remain underserved by financial markets in many emerging and developing economies. In a well functioning financial system, SMEs are more likely to have a range of financing options and support services as they grow. A "typical" small business start-up in the United States, for instance, is likely to have access to personal savings and contributions from friends and family to finance the initial launch with additional funding from angel or venture capital investors as well as traditional bank loans for the latter stages of the business development cycle. In contrast, SMEs in developing countries typically operate in a much less supportive environment. For sustainable SMEs, the organizational barriers are even higher. Banks are particularly reluctant to support businesses in rural areas, where many sustainable SMEs are located, and they tend to be very cautious about lending in relatively new product and industrial sectors such as organic farming or renewable energy generation (Barreiro, Hussels, and Richards 2009).

If SMEs are unable to raise capital to take advantage of opportunities to design new products and services, then they are less likely to become larger and more successful companies. This situation is what many economic development experts refer to as the "missing middle" dilemma, where SMEs are caught in the middle of the business financing cycle—too large to access microcredit financing mechanisms and too small, unstable in terms of cash flow, and lacking the required collateral to attract longer-term growth financing from commercial banks (Yago, Roveda, and White 2007). Many multinational corporations like Microsoft, eBay, and Google were all once SMEs and received many rounds of venture capital and other financing before becoming world-renowned multinational business enterprises.

While the need for funds for sustainability-oriented SMEs in developing and emerging economies always exceeds the available capital base, there are a number of impact investment funds focused on achieving environmental and social objectives as well as providing technical assistance in developing countries. Examples of such organizations include: the New Jersey-based E + Co that invests in sustainable energy SMEs; Cambridge, Massachusetts-based Root Capital that provides financing for grassroots businesses in rural areas of developing countries; Desi Power (India) and Lotus Energy (Nepal) that support a wide range of energy, environmental, and social microenterprises; and EcoE that focuses on the small but growing green business sector in Latin America. EcoE is examined further in the case study that follows.

EcoEnterprises Fund Case Study

EcoE is part of a pioneering movement to harness business to effect environmental change. For over a decade, EcoE has supported the growth of sustainable enterprises in Latin America to demonstrate how economic viability can bring about positive conservation impacts. EcoE offers risk capital and business advisory services to small-scale and community-based environmentally and socially responsible businesses involved in ecotourism, sustainable agriculture, apiculture, sustainable forestry, and nontimber forest products. The original seed money for EcoE was provided by the Multilateral Investment Fund of the Inter-American Development Bank, The Nature Conservancy, Corporación Andina de Fomento, International Finance Corporation, foundations, and private investors. The Nature Conservancy serves as a fund manager, while a board of directors comprised of leaders from conservation, business, and investment fields provides oversight.

Between 2000 and 2009, EcoE invested $6.3 million in 23 companies across 10 Latin American countries (see Table 4.1 for a partial list of EcoE's investments). During this period, companies leveraged an additional $138 million in investment capital and generated $281 million in sales, providing economic benefits for local economies. Positive environmental impacts through habitat restoration, ecosystem protection, and biological corridor preservation resulted as well. The companies directly conserved 2.1 million acres (860,773 hectares) in 19 biologically significant areas, 7 of which are World Heritage Sites.

EcoE invests in innovators in their respective industry sectors. In addition to Sambazon, which introduced açaí to the marketplace in the United States and Europe, EcoE invested in BioCentinela, one of the world's first certified organic shrimp operations. BioCentinela is managed by a lawyer-turned-entrepreneur and is steadily restoring the critical mangrove habitat of Ecuador while selling organic shrimp products. There are now over 70 different animal species at the project site, including several rare birds. Other EcoE portfolio

Table 4.1 Selected companies in EcoEnterprises' fund investment portfolio

Ecotourism
- Belize Lodge and Excursions (http://www.belizelodge.com)
- Rolf Wittmer Turismo Galápagos Cia. Ltda. (http://www.rwittmer.com)
- Rainforest Expeditions (http://www.perunature.com)
- Veragua Rainforest (http://www.veraguarainforest.com)

Non-Timber Forest Products
- Sambazon do Brasil Representação Comercial Ltda. (http://www.sambazon.com)

Sustainable Agriculture
- Orgánico SR Los Nacientes, S.A.
- ForesTrade de Guatemala, S.A. (http://www.forestrade.com)
- Loofah, S.A. (http://www.loofahshop.com)
- Rainforest Exquisite Products, S.A. (http://www.rainforest-products.com)
- Organic Blooming, S.C.C. (http://www.organiccallas.com)
- Terrafertil, S.A. (http://www.terra-fertil.com)

Sustainable Aquaculture
- 8th Sea – The Organic Seafood Company
- Biocentinela, S.A. (http://www.biocentinela.com)
- Aqua Consult International, S.A. de C.V.
- Marimex del Pacifico, S.A. de C.V.

Sustainable Forestry
- Jolyka Bolivia, S.R.L. (http://www.jolyka.com)
- Interforest, Ltd. (http://www.interforest.com.gt)
- Suma Pacha Industriales de Bolivia, S.A. (http://www.forestworld.com)
- Noram de Mexico, S.A. de C.V.

companies include a manufacturer of garden furniture that sources its wood from sustainably managed forests in Bolivia, a biological and research adventure park in Costa Rica, a farm to produce algae in Mexico, and a chain of ecolodges in Peru. Even though EcoE invested during a difficult economic decade, 20 of the 23 businesses it has invested in are still in operation. These successful enterprises demonstrate that SMEs with access to capital can successfully overcome the challenges faced and achieve Triple Bottom Line results.

Key Lessons To Be Learned

EcoE's experiences with sustainable SMEs suggest a series of possible lessons that might be learned by other organizations working in similar situations.

Partnering: Community Collaboration Pivotal for Success

One important lesson learned by EcoE is that environmental nongovernment organizations (NGOs) and local community groups play a crucial function in the commercial viability of an environmental enterprise. Community partnerships and strong stakeholder relations must be an integral component of a sustainability-committed company's approach and organizational philosophy. NGO-private-sector collaborations can consist of informal as well as formal arrangements, including purchase agreements with producer groups that serve as a critical part of a company's supply chain; outreach with local stakeholders who live near project sites and whose support must be earned to ensure smooth operations; and cooperation with NGOs to provide environmental and social monitoring, training or research, and other services. Active participation of all local stakeholders is inherently linked to solid biodiversity outcomes and sustainable natural resource management. In total, EcoE investees partnered with 293 community groups and NGOs, generated 3,513 jobs, and worked with more than 13,231 local producers. An estimated 98,780 local people throughout Latin America have benefited from these company-community engagements.

Differing Agendas: Finding Common Objectives

Businesses and local stakeholders can have differing agendas, but need to have common objectives. There is an inevitable "dance" between the company, NGO, local community, and/or other vested parties. Beneficial outcomes require hard work and a strong commitment to communication, negotiation, and dialogue. Unlike conventional businesses, EcoE's investees have social and environmental mandates to guide their overall organizational strategies. But despite these values, things such as uncomfortable situations, sabotage, and/or lawsuits can occur if relationships are not well managed. Trust must be established and can be accomplished by identifying all issues early on and incorporating all stakeholders in the business planning process. For instance, the community needs to have an understanding of company value-added in the marketplace to uphold the integrity of its product. For one EcoE portfolio company that sells organically certified products and fosters biodiversity standards as part of their branding platform, it was discovered that a community that committed to the company's sustainability principles was cutting down part of the forest to increase output. The company was forced to drop the community as a supplier, to the community's surprise. However, the company then worked with local leaders to ensure that its standards were again being met, and the supplier relationship was reestablished in the following year. Peru-based ecotourism company Rainforest Expeditions and the local Infierno community established joint equity arrangements at the start of their joint venture and then built other income-generating activities for community members beyond the community's stake in the lodge. Sustainable forestry company Noram, based in the Sierra Madre Mountains of Mexico, has been able to survive business fluctuations by relying on the local *ejidos* (indigenous community organizations) that supply the company with its raw materials. The ejidos-Noram relationship was built over years through Noram's substantive support of forest certification efforts in the region.

Managing Expectations: Communication and Follow-Through Can Not Be Underestimated

The balancing of expectations among interested parties can be upset by negative experiences (e.g., unkept promises, contract breaches, relationship ambiguities, and power struggles). Consequently, it is critically important for partnerships of business, NGO, and local community to lay out a practical plan of action with measurable steps that can be evaluated through ongoing coordination and communication. Giving voice to the local population, especially women or indigenous peoples, as part of the process contributes to the long-term prospects for the working collaboration. Companies such as Sambazon and Noram learned early on that providing immediate cash for product was an essential element to successfully establishing a stable raw material supply from local organizations. In cooperation with community leaders, Sambazon, Rainforest Expeditions, and Marimex, a Mexican sustainable aquaculture company, have expanded community engagement in order to fortify the sustainability of operations. Marimex has long realized that its business depends on sustainable management practices of the local community members who live near its operations. The company therefore provides the appropriate training and technology to strengthen the institutional capabilities of its community partners.

Leadership: In the End, It's All About Management

Undoubtedly, it takes a unique entrepreneur to handle all of the complexities involved in launching and managing a Triple Bottom Line business in rural Latin America. For a conservation-oriented business to succeed, a commitment to social impact and community engagement has to be an essential feature of a company's business strategy. This commitment may take the form of training, providing medical services, building a well for fresh water, or offering educational opportunities. At the same time, the local community needs a leader to pursue the best interests of the local peoples while having an understanding of what will and will not work with and for the company. Community leaders who are solution-oriented when addressing the community's needs and who speak on behalf of the entire community are most able to devise an effective, sustainable, and long-lasting arrangement. Such leadership is indispensable for indigenous communities given their own social structure and generally limited past engagement in economic activity outside of the community. EcoE's investees have benefited from a number of enlightened community and NGO leaders who had the wisdom and foresight to negotiate workable partnerships in which all vested parties gained. For instance, after several attempts to work effectively with various communities, Sambazon teamed up with NGO Peabiru to guide its approaches, and recently established the Sustainable Amazon Partnership in cooperation with this and other local groups.

Adapting with Focus: Keeping the Integrity of the Business Model

Another important lesson from the EcoE case study is that to find success, a company needs to have a resilient strategy that adapts to market realities, without moving away from the authenticity of the underlying business concept. EcoE's portfolio companies pioneer novel opportunities; for an SME this is a risky and expensive venture. Nonetheless, the entrepreneurs see themselves as change agents bringing about new ways of doing business while showing the potential for realizing an integrated environmental, social, and economic return. This value-added commitment is a critical contribution to the evolution of the industry sector, investor interest, and consumer demand for their niche products.

First Mover Challenges: Being Market Pioneers Is Exciting, but Risky

EcoE portfolio companies tend to be "first movers" in their respective industries. The founders of these companies are usually entrepreneurs with a great passion, and their initial success comes from their ability to sell the concept on which their new venture is based. These companies, located throughout Latin America, have often been leaders paving the way for other companies to enter the market. Being early movers, many EcoE portfolio companies bear the burden of dedicating significant amounts of their invaluable time and limited capital to training, research, market development, and consumer education. Developing a new product and then taking it to market requires a significant investment of resources—no doubt hard to come by for start-up companies. Considering these companies are launched in rural poor areas, rely on community partnerships, have underlying environmental and social values mandates, and sell product in the export market, the hurdles are awesome.

When the market takes off, there are significant prospects at hand for these companies, but they may have inadequate access to the longer-term capital needed for growth. In many cases, the company may not have sufficient financial and organizational resources to wait until its investments in market development start to pay off. EcoE's first investment in organic cocoa did not succeed because it was an early entrant in the marketplace and the additional capital needed to take the company to the next level fell through at a critical juncture. Years later, the same entrepreneur tried again and was able to capitalize on the explosion in the demand for organic chocolate. There may also be unforeseen events that impact company performance. Downturns in the global economy undoubtedly affect ecotourism travel although this specific segment in the tourism sector is more resilient. EcoE's investee, the Bolivian sustainable forestry company Jolyka, was able to recover from the September 11 (2001) downturn in construction by diversifying and solidifying its presence in the European market.

Persistence: Slow and Steady May Win the Race

Rapid business growth often requires large reserves of cash. Some companies pursue strategies to scale-up quickly using available financial resources with the risk that the company may have only limited funding for latter stage business operations. Many EcoE portfolio companies, including the Ecuador-based certified organic cut flower producer and exporter Organic Blooming and one of Mexico's first algae companies Aqua Consult International, take a different approach. They entered their respective markets gradually and achieved small successes while improving technical issues and developing the market. Having an angel investor is a significant factor in adopting this slow and steady course of action. Angel investors represent a key element in the success of start-up companies and first movers by providing "patient" investment capital that allows the company to be more financially and organizationally secure. With its headquarters in the United States, Sambazon was able to find and benefit from an angel investor. Such an opportunity is unusual for SMEs in Latin America because of the lack of angel networks in the region.

Networks and Experience: Important Business Assets

Undoubtedly, the sustainable entrepreneur's background is vitally important in bringing a new product to market. In EcoE's portfolio, Ecuador-based organic dried and fresh fruit company Terrafertil is a good example. The principal of this company acquired years of experience in a family food packaging business, and it is through this company that Terrafertil had access to Ecuador's largest local supermarket chain. Terrafertil was then able to generate cash flow and enhance its product and company capabilities as it worked on export sales. Terrafertil also invested heavily in research, planning, and new product and market development, which provided a solid foundation for growth.

Other EcoE investees followed similar steps and, in cases such as those of Organic Blooming, Marimex, and BioCentinela, sought advice from technical experts in the organic agriculture or aquaculture fields to ensure technical and quality performance. As is the case in many start-up companies in the United States and other industrialized countries, entrepreneurs that are so important to the early success of their companies may lack the financial and business skills required to manage the next stage of the company's strategic development. There were a few EcoE investees in which new management was hired. But, in Latin America where most SMEs are family-owned, bringing in new talent and senior management is unusual, and this fact is reflected in EcoE's portfolio.

Conclusion

What constitutes business or market success? For conventional investments and entrepreneurial ventures, the answers to this question are straightforward. Money raised and financial returns achieved are often offered as answers. Undergoing an initial public offering or achieving some revenue and/or asset benchmark indicates a win. However, for EcoE's investment niche—sustainable entrepreneurs and business ventures seeking social, environmental, and economic benefits—defining success is inherently more complex. While those involved in the theory and practice of sustainable entrepreneurship and investment issues are starting to design, develop, and implement performance metrics that go beyond purely financial ones as well as experimenting with Triple Bottom Line evaluation techniques, the difficulties of measuring nonfinancial returns continue to be barriers to achieving a simple and understandable formula for evaluating success.

Recent establishment of the Aspen Network for Development Entrepreneur's Social Metric project (ANDE 2010), the Global Impact Investment Network's Impact Reporting and Investment Standards project (GIIN 2010), and other related initiatives in the past couple of years reflect a growing international interest and effort to develop a Triple Bottom Line value-assessment methodology for the larger global investment and business community. Yet, just as the success of the Global Reporting Initiative standard did not replace the conventional corporate financial reporting framework, it is likely that the measurement guidelines these efforts are developing will be a supplemental influence on mainstreaming this investment approach. The primary way sustainable entrepreneurship and investment will gain a foothold in the global finance and business landscape is through learning from illustrative examples and responding strategically to the evolving marketplace.

What has been happening in EcoE's portfolio supports this view. EcoE's proof-of-concept phase over the past ten years has led to the design of the next generation fund, EcoE II, with an anticipated capitalization of $30 million, six times greater than the first fund. It is clear that there are many and increasingly stronger investment opportunities in the sustainable entrepreneurship and investment sectors. The continued devastation of critical habitats in biodiversity- and ecosystem-rich regions has increased the need for businesses that can use local natural resources wisely and fuel long-term economic growth. Markets for organic food, renewable energy, and products sustainably harvested from forests and oceans continue to grow, while concern over global climate change has generated an economic interest in activities that help mitigate environmental impacts and offset carbon emissions. As Ryan and Jeremy Black of Sambazon discovered, sustainable entrepreneurship with the investment capital and support from such investors as EcoE can yield profits and provide benefits to local communities and planet earth.

References

Anderson, T. L., and D. R. Leal. 1997. *Enviro-capitalists: Doing good while doing well.* Boston, MA: Rowman & Littlefield.

Aspen Network for Development Entrepreneur (ANDE) 2010.

http://www.aspeninstitute.org/policy-work/aspen-network-development-entrepreneurs (accessed April 24, 2010).

Austin J., H. Leonard, E. Reficco, and J. Wei-Skillern. 2006. Corporate social entrepreneurship: The new frontier. In *The accountable corporation. Volume 3: Corporate social responsibility,* ed. M. Epstein and K. Hanson. Westport, CT: Praeger.

Austin, J., H. Stevenson, and J. Wei-Skillern. 2006. Social and commercial entrepreneurship: Same, different, or both? *Entrepreneurship: Theory and Practice* 30(1): 1–22.

Barreiro, V., M. Hussels, and B. Richards. 2009. *On the frontiers of finance: Scaling up investments in sustainable small and medium enterprises in developing countries.* Washington, D.C.: World Resources Institute.

Bielak, D. B., S. Bonini, and J. M. Oppenheim. 2007. CEOs on strategy and social issues. *McKinsey Quarterly* (October): 1–8.

Borstein, D. 2004. *How to change the world: Social entrepreneurs and the power of new ideas.* Oxford, UK: Oxford University Press.

Chatterji, A., D. Levine, and M. Toffel. 2007. How well do social ratings actually measure corporate social responsibility? Working Paper No. 33 Cambridge, MA: Kennedy School of Government, Harvard University.

Dean, T., and J. McMullen. 2007. Toward a theory of sustainable entrepreneurship: Reducing environmental degradation through entrepreneurial action. *Journal of BusinessVenturing* 22(1): 50–76.

Drucker, P. 1985. *Innovation and entrepreneurship.* Oxford, UK: Butterworth-Heinemann.

EcoEnterprises Fund. http://www.ecoenterprisesfund.com/portfolio.html (accessed April 24, 2010).

Frankental, P., and F. House. 2000 *Human rights—is it any of your business?* London: Amnesty International UK and the Prince of Wales Business Leaders Forum.

Freireich, J., and K. Fulton. 2009. *Investing for social and environmental impact: Design for catalyzing an emerging industry.* Boston, MA: Monitor Group.

Global Impact Investing Network (GIIN). 2010. http://www.globalimpactinvesting network.org (accessed April 24, 2010).

Haufler, V. 2006. Global Finance, Power and Social Purpose. Paper presented at the Annual Meeting of the International Studies Association (March 21–26). 2006.

Hill, R. P., T. Ainscough, T. Shank, and D. Manullang. 2007. Corporate social responsibility and socially responsible investing: A global perspective. *Journal of Business Ethics* 70(2): 165–174.

Ivanko, J. D, and L. Kivirist. 2008. *Ecopreneuring: Putting purpose and the planet before profits.* Gabriola Island, BC: New Society.

Larson, A. 2000. Sustainable innovation through an entrepreneurship lens. *Business Strategy and the Environment* 9: 304–317.

MacLeod, M., and J. Park. Forthcoming. Financial activism and global climate change: The rise of investor-driven governance networks. *Global Environmental Politics.*

Mair, J., and K. Ganly. 2010. Social entrepreneurs: Innovating toward sustainability. In *State of the world 2010: Transforming cultures: From consumerism to sustainability,* ed. J. Mair and K. Ganly (pp 103–109). New York and London: W. W. Norton & Company, 103–109.

Margolis, J., and J. Walsh. 2001. *People and profits?: The search for a link between a company's social and financial performance.* Mahwah, NJ: Lawrence Erlbaum Associates.

Mercer. 2009. *Shedding light on responsible investment: Approaches, returns, and impacts.* Toronto: Mercer Investment Consulting.

———. 2006. *Perspectives on responsible investing.* Toronto: Mercer Investment Consulting.

Newmark, T. 2009. Green venture capital. *Americas Quarterly* (Fall): 96–97.

Park J. 2004. Business and human rights: Mixing profit and principles in the global marketplace. *Journal of Corporate Citizenship* 13: 24–27.

———. Forthcoming. Responsible investing: Challenges and opportunities for sustainable strategic management in the global context. *International Journal of Sustainable Strategic Management.*

Parthiban, D., M. Bloom, and A. Hillman. 2007. Investor activism, managerial responsiveness, and corporate social performance. *Strategic Management Journal* 28(1): 91–100.

Pralahad, C. K. 2004. *The fortune at the bottom of the pyramid: Eradicating poverty through profits.* Philadelphia: Wharton School Publishing.

Schaltegger, S., and H. Petersen. 2001. *Ecopreneurship: Konzept und Typologie.* Lüneburg/Luzern, Germany: Center for Sustainability Management.

Schaper, M. 2002. The challenge of environmental responsibility and sustainable development: Implications for SME and entrepreneurship academics. In *Radical changes in the world: Will SMEs soar or crash?* Ed. U. Füglistaller, H. J. Pleitner, T. Volery, and W. Weber (pp. 525–534). St. Gallen, Switzerland: Rencontres de St. Gallen.

Schaper, M. ed. 2005. *Making ecopreneurs: Developing sustainable entrepreneurship.* Surrey, UK: Ashgate Publishing.

Schumpeter, J. A. 1934. *The theory of economic development.* New Brunswick, NJ: Transaction.

———. 1950. *Capitalism, socialism and democracy.* New York: Harper and Row.

SustainAbility 2008. *The social intrapreneur: A field guide for corporate changemakers.* London: SustainAbility.

UN Economic and Social Commission for Asia and the Pacific (UNESCAP). 2005. *Energy services for sustainable development in rural areas in Asia and Pacific policy and practice.* Energy Resources Development Series No. 40, Bangkok: UNESCAP.

Vernon, R. 1977 *Storm over the multinationals: The real issues.* Cambridge, MA: Harvard University Press.

World Resources Institute, UN Development Program, UN Environment Program, and the World Bank. 2008. *World resources report: Roots of resilience.* Washington, D.C.: World Resources Institute.

Yago, G., D. Roveda, and J. White. 2007 *Transatlantic innovations in affordable capital for small and medium-sized enterprises: Prospects for market-based development finance.* Washington, D.C.: German Marshall Fund of the U.S.

PART II

Action Implications: Motivating and Influencing Business

CHAPTER 5

Sustainability in Business Curricula: Toward a Deeper Understanding and Implementation

Blanca Luisa Delgado-Márquez, J. Alberto Aragón-Correa, and Nuria E. Hurtado-Torres

Introduction

The need to integrate Education for Global Sustainability (EGS) into all levels of the educational system has taken on a new priority in recent years. The aim of sustainability education is to equip students with leadership skills, management capabilities, and the broad knowledge needed to create new approaches that can lead to global sustainability.

The links between education and global sustainability are perceived as being very strong and, as documented by the World Commission on Environment and Development (WCED 1987), by Agenda 21 (UN 2002), and by recent environmental reports (UNEP 2007), greater efforts are especially important to guarantee the integration of sustainability into university education. However, in a study published in 2004, Wright concluded that the initiatives to promote the concept of sustainability in higher education had actually had little impact on university courses.

Some universities concerned with environmental issues have developed a variety of initiatives for integration of sustainability into their curricula. Some of these cases have been reported in the literature (e.g., Aspen Institute 2009; Sammalisto and Arvidsson 2005). The diversity in objectives and approaches is very high, although all approaches emphasize the importance of moving in the direction of a greater introduction of sustainable contents into university education.

The central questions of how and why universities decide to integrate sustainable contents into their courses remain unexplored. Given the importance of management students acquiring skills related to sustainability—skills that can be used in their professional lives as managers to ensure a more sustainable world—it is important to define ways of influencing and increasing the integration of sustainable contents into management education.

To contribute to this goal, this chapter is divided into three parts. The first part presents a review of the importance of EGS in universities. In the second part of the chapter, we analyze, from an international perspective, the current offerings of sustainable MBAs, and we review the latest literature about the integration of sustainability into graduate business schools. This part of the chapter illustrates the aspects of business education that can be labeled "sustainability-oriented" and describes the scarcity of these aspects in MBA programs. In the third part of the chapter, we report on an empirical study of business schools in Spain, where the level of environmental development can be considered to be "medium" in comparison with other Western countries. Because the Spanish government has promoted the Bologna process for "modernization" and a higher integration of the Spanish university system in the approved European approach, and because Spain is in the middle of the process of changing its university system, a detailed analysis of the Spanish situation offers an interesting illustration of this change process and provides a good baseline for comparing with past, future, and other situations.

Education for Global Sustainability in Universities

Literature analyzing the integration of sustainability issues into university education has focused on proposing pedagogical teaching tools for educators (Porter and Córdoba 2009), case studies on the integration of sustainability in some MBA courses (Benn and Dunphy 2009), or approaches to developing leaders capable of ensuring sustainability (Hind, Wilson, and Lenssen 2009).

The possibilities for integrating contents related to sustainability are multiple and include the existence of one or more specific topics within a course, specialized courses, or even specific programs for sustainable management. Courses in business schools, however, are still far from providing comprehensive philosophies or contents aimed at helping to understand the importance of sustainability in the technical knowledge of business management (Hart 2008).

A study by Diane Holt (2003) favored integrating discipline-specific modules into other subjects. The study observed that "when environmental modules are integrated in corporate and management classes" (p. 329), they might be more effective than specific courses designed "only" to increase awareness of sustainability. This statement, based on a study of the "impact of education and cultural experiences of business school students during their 3 years at the University" (p. 331), supports the idea of integration rather than separate courses. This finding is contrary to the practices in many universities, where the focus has been on establishing general environmental courses, thus providing an overview of environmental problems, concepts, and approaches. Therefore, it is crucial to link sustainability concepts to other disciplines and topics rather than providing isolated sustainability-oriented courses.

A number of studies have proposed models for integrating sustainability into education (e.g., Sterling 2004). Many of these studies are summarized by Sammalisto and Lindhqvist (2008), who proposed that three steps may be taken to integrate sustainability into education and that it would be appropriate to go as soon as possible to the highest levels. The first level of integration of sustainability is called "bolting-on," i.e., adding the concept of sustainability to the existing curricula and educational system, which in itself remains largely unchanged. This approach is "education about sustainability" and can consist of separate courses about sustainability.

The second level of integration, "building-in," involves a deeper level of response. It implies that ideas are incorporated into existing educational systems, e.g., greening the curriculum and institutional operations. This level could be called "education for sustainability," and it includes integrating sustainability issues into regular discipline-specific courses. It aims at creating a connection in the minds of students between the subject in question and sustainable development. Some of the earliest initiatives in Sweden focused on technical and economic programs as well as on teacher training programs and "were carried out as projects sponsored by the National Council for Renewal of Higher Education. The results from projects in 27 universities indicated a positive impact on students, particularly for engineering students in the smaller universities" (Sammalisto and Lindhqvist 2008: p. 222).

The greening of engineering programs was spread over a larger number of subjects as compared with programs in the discipline of economics, in which greening was limited to the core subjects (Sammalisto 1999). An approach somewhat similar to the Swedish one has been applied in The Netherlands, based on disciplinary reviews. This approach poses an intellectual challenge to instructors to integrate sustainability within each discipline by "exploring the relationships between various disciplines and sustainable development" (Appel, Dankelman, and Kuipers 2004).

The third level is called "transformation," a term that implies a complete redesign of education based on sustainability principles. This level would require a paradigm change so that education would be built on learning as change and education as sustainability. In practice, the transformation level would mean that the goal of all education would be to teach sustainable development, and the different disciplines and subjects would all contribute to it. An example of a model for integration to this extent is the one presented by Juárez-Nájera, Dieleman, and Turpin-Marion (2006). They presented a framework for a culture where holistic understanding is the focus of education rather than specific knowledge or skills: "Students must learn new and sustainable ways of looking at the world, themselves and their professions" (p. 1037).

Because very high levels of commitment and innovation would be required to take the transformational approach and many attempts to break such new intellectual and professional ground might fail, we agree with Holt (2003) and Sterling (2004) that integrating sustainability into normal courses and research projects at a university is currently the most promising way of reaching the students and equipping them with effective tools and knowledge to work for sustainability in their future careers. For this reason, we studied the extent to which the future integration of sustainability into management courses depends on the preferences of department heads. In doing so, we paid special attention to the incorporation of environmental contents as a complement to the economic, and even social, contents traditionally addressed in management courses.

It is also important to recognize that internal and external barriers can hinder the integration of sustainability into the normal curriculum. Different studies have identified problems in the process of integration, such as lack of the highest administrators' commitment to the work of integrating sustainability, lack of follow-up procedures, failure to recognize and accept the importance of environmental and sustainability problems, limited time and resources, and the prevailing academic culture (Lidgren, Rodhe, and Huisingh 2006; Sammalisto and Arvidsson 2005; Segalàs, Cruz, and Mulder 2004).

The Sustainable MBA and the Business Schools: An International Perspective

Some business schools are heeding the call for incorporating greening and sustainability into their academic offerings in increasingly meaningful ways. They are integrating the topic into existing courses or subject areas, addressing both theoretical and pedagogical perspectives. Several management disciplines have already implemented sustainable contents into business curricula, such as operations management, strategic management, human resources management, business and society, and international management.

A traditional Master in Business Administration (MBA) degree is focused on an orientation of business for financial profits. The main courses are usually related to financial issues, with some attention to management, marketing, or economic theory, among others. Societal, environmental, or ethical implications of managerial decisions are often absent in major courses or are just marginal in some minor courses. In contrast, "sustainable MBA programs" have been proposed to add the study of managing for environmental and social sustainability to these subjects. Sustainability in these programs is generally understood as the inclusion of economic, environmental, and social sustainability, collectively known as the Triple Bottom Line (Elkington 1997). For each of these domains, the approach to teaching sustainability implies that taking the appropriate actions will make it possible to continue through the near future without economic, environmental, or social breakdowns.

Several schools have already implemented "green" or sustainable MBA programs. Examples include the University of British Columbia (MBA—Sustainability and Business Specialisation), Marlboro College (MBA in Managing for Sustainability), Colorado State University (Global Social and Sustainable Enterprise Program), the Bambridge Graduate Institute (MBA in Sustainable Business), and others.

Aspen Institute's initiative is probably the most famous and widespread action designed to promote sustainable MBAs. Its initiatives include a research survey and alternative ranking of business schools that spotlights innovative full-time MBA programs integrating issues concerning social and environmental contents into the curriculum. Figure 5.1 presents the percentage of sustainable MBAs available in each country. The vast majority of green MBAs are located in universities in the United States (65.33 percent). A second group of countries is formed by Canada (6.67 percent) and United Kingdom (5.33 percent). The remaining countries offering sustainable MBAs are far from providing a large number of green MBA programs.

Figure 5.2 gathers the absolute number of sustainable MBA programs offered by the same countries. The United States holds the dominant position in this area, with a total of 98 green MBAs. Canada offers 10 of the 150 sustainable MBAs available in the world, and the United Kingdom offers 8. Although Aspen is the most successful initiative to promote sustainable MBAs, the inclusion of 150 MBA programs is still marginal compared with the vast number of MBA programs in the world.

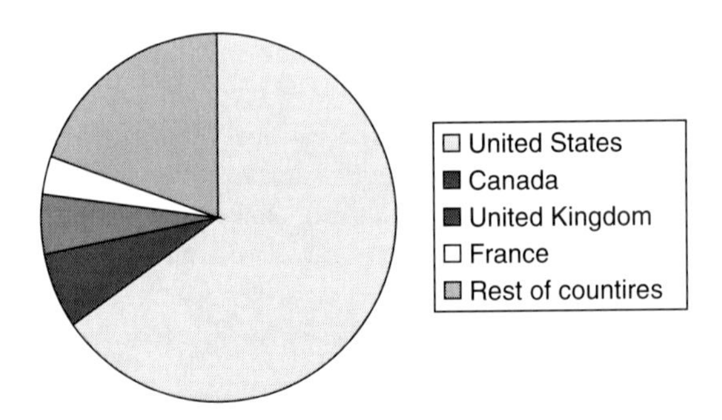

Figure 5.1 Percentage of sustainable MBAs per country
Source: Self-elaboration with data extracted from Aspen's Institute

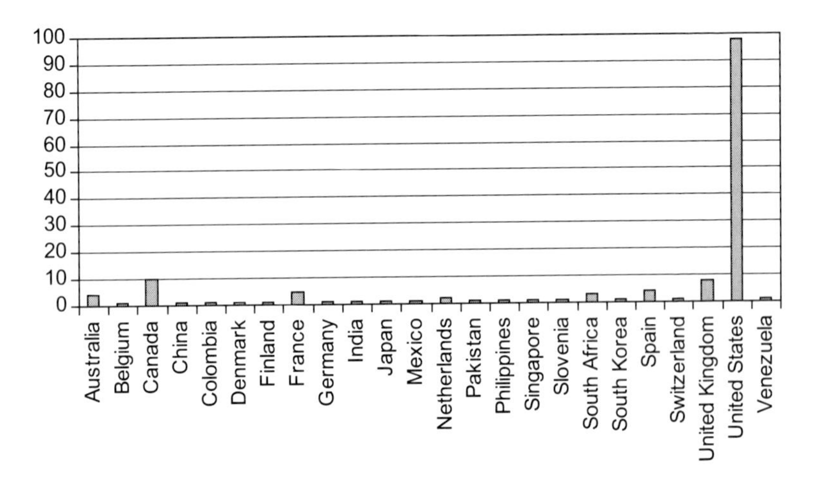

Figure 5.2 Number of sustainable MBAs per country
Source: Self-elaboration with data extracted from Aspen's Institute

Recent literature has described an increasing interest for the integration of sustainability into management education (e.g., Wright 2004). However, most business schools are out of step with community attitudes and the expectations of many business leaders concerning sustainability and corporate responsibility—they are lagging rather than leading (Starkey and Welford 2001). Walck (2009) points out, from a dean's perspective, that the integration of sustainability and environmental ethics into management education responds, among other factors, to internal pressure from faculty who have focused their research on the intersection of business and the natural environment. Steketee (2009), describing the continuing evolution of Aquinas

College's four-year Bachelor of Science program in sustainable business, further emphasizes the connectivity among business leaders, grant makers, academic faculty, and students.

It is important to highlight that scholars also face challenges related to the integration of sustainability into management education. Moreover, Peoples (2009) claims that business leaders and their academic trainers must embrace the concept of sustainability to prepare future leaders with the understanding and tools necessary to make key decisions based on more than "just the numbers."

In the international sphere, a recent study is useful to show how sustainability is integrated across the MBA curriculum at a university in Australia and what some of the teaching challenges are (Benn and Dunphy 2009). Findings indicate there is substantial variation in student attitudes to sustainability. While some students have already developed high levels of interest in the area and actively welcome the introduction of sustainability-related contents, many other students have very traditional ideas of what an MBA is about, so it is important to emphasize the business case for sustainability. Moreover, they point out that how-to cases that stress the importance of pragmatic skills are more appealing to many MBA students and to faculty alike than why-cases that deal with theoretical or ideological issues.

In other recent teaching approaches, Walker et al. 2009 have addressed sustainable procurement in the health sector in a virtual classroom setting, and Gundlach and Zivnuska (2010) found the use of an experiential learning approach to be a more effective teaching tool than a traditional lecture approach when teaching social entrepreneurship and sustainability.

Rands (2009) introduces and extends a principle attribute matrix (PAM), which is a flexible tool to help instructors integrate environmental sustainability into management courses at both undergraduate and graduate levels. Use of the PAM can help students learn and apply environmental sustainability concepts in their managerial and personal lives. It also allows instructors to choose the extent to which they wish to integrate environmental sustainability in their courses—for example, with one or two topics, a module, or the entire course. Rands argues for the importance of enhancing traditional course materials such as readings and cases with applied learning and subsequent reflection when delivering greening lessons.

Recent studies apply systems theory to sustainability education. Porter and Córdoba (2009) offer three general approaches to systems thinking: functionalist, interpretive, and complex adaptive systems. In situations where there are predefined goals about sustainability or identified needs for improvement, the use of functionalist system methods can be of great help. In other contexts where engagement is preferred and where sustainability is seen as a process, the interpretive framework is more suitable. Finally, they state that in cases where the complexity of issues, multiplicity of perspectives, relevant levels of analysis, and desire to involve a wide range of constituents in change efforts are all high, the best approach may be that of complex adaptive systems.

Finally, Bradfield (2009) argues that sustainability innovation is fundamentally different from classic process improvement because sustainable innovations force people to change the way they have always done things. Curricula should therefore teach students how they can function as entrepreneurs inside a large corporation or a small company with limited capital.

Current Situation and Future Outlook of Sustainability in Business Curricula: The Case of Spain

European universities are developing a full process aimed at achieving a greater homogeneity of their national university systems. The "Bologna process" was designed to develop a common approach in university education for countries integrated in the European Union. The process, however, is also a clear opportunity to ensure that Spanish university curricula are updated, including aspects of sustainable development as part of the comprehensive education of students. The EHEA (European Higher Education Area) will involve substantial changes in syllabi, the emergence, deletions and/or fusion of careers, as well as conceptual and methodological reformulations.

As mentioned, Spain is a country whose level of environmental development can be considered as medium. Although Spain is one of the European countries with a larger number of certified firms with ISO 14001 and EMAS, the European Union statistics show that its population is only marginally interested in the consumption of ecological products. The Spanish government, beyond promoting the Bologna process, has also approved new legislation for a "modernization" of the Spanish university system (see Organic Law 4/2007 on Universities of 12 April, and Royal Decree 1393/2007 of 29 October). The law states: "It should be noted that training in any professional activity should contribute to knowledge and development of human rights, democratic principles, equality between women and men, solidarity, environmental protection, accessibility and universal design for all, as well as promoting culture of peace."

A detailed analysis of the Spanish situation provides an interesting illustration of the situation in an environmentally average European country.

Some Spanish institutions, such as the Polytechnic University of Catalonia, are pioneering institutions where the syllabi of different disciplines introduce case studies and teaching methodologies for sustainability training of their students (Aznar and Ulls 2009).

A detailed study of the extent to which sustainable contents are integrated into the business curricula in Spain has not yet been conducted. Therefore, considering this situation, our attention will be focused on answering the following four questions:

1. What specific sustainable topics have higher importance in business schools nowadays?
2. What specific sustainable topics are expected to have higher importance in business schools in the future?
3. How does the head of a department perceive demands of society regarding the incorporation of sustainable topics?
4. Does the forecast of the likelihood of integrating sustainability issues into the curriculum in the future differ based on the head of department's preferences to integrate sustainable contents?

To answer these questions, we used a sample comprised of department heads of management and business schools from 25 Spanish universities. The sample was obtained by means of a written questionnaire sent on three occasions by both regular mail and by e-mail to the heads of departments, with the option to reply online or on paper. Our final sample consisted of 58 heads of departments. It is important to note that department managers in Spain are largely responsible for the specific syllabus and orientation of the courses, whereas other administrators (e.g., deans) play a broader role, coordinating departments and school management, and serve as symbolic representatives. Here, we describe the results of our investigation.

Current Situation and Forecast

To assess the sustainability contents in the current curricula in management, we measured the extent to which seven issues are integrated into the departments' curricula. Those issues are "environmental regulation," "environmental management systems and environmental certification," "corporate social responsibility and ethics," "environmental implications of operations and wastes," "energy saving," "sustainable economics," and "design of technology respectful with the natural environment." We used a Likert five-point scale ranging from "this issue is not in the course at all" to "this issue is a key for all the department's courses" (see Figure 5.3).

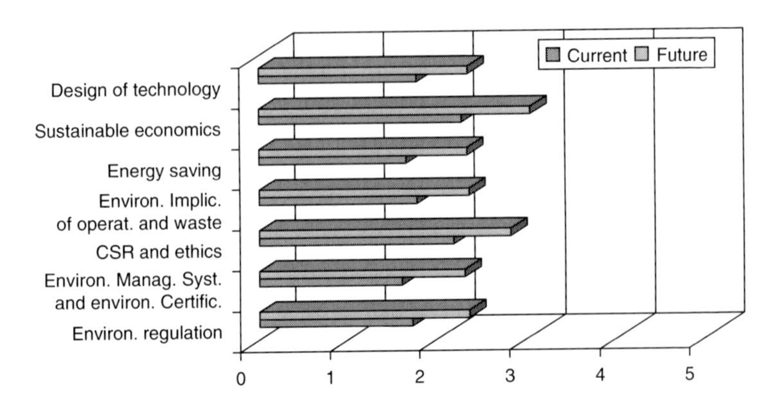

Figure 5.3 Current and future sustainable contents

In relation to forecasts about the integration of sustainability issues into business courses in the future, we asked respondents to describe their prevision regarding the future integration of the seven items described earlier. This prevision was measured by using a Likert scale from 1 to 5, in which higher values indicated more advanced intentions regarding the integration of environmental topics.

The results show that almost all contents that could provide environmental skills to graduates are treated in a reduced way, receiving only residual attention in the current curricula. As for future preferences, Figure 5.3 shows that they are substantially higher compared to the current situation. The topics where the greatest future development is expected are "sustainable economics" and "corporate social responsibility and ethics." These issues are also receiving greater attention in current curricula.

Figure 5.3 suggests that sustainable contents will receive somewhat greater attention in future curricula, but that future attention will still not be very large. How much can society and the various university groups demand that universities incorporate more sustainable contents? To assess the interest of the different groups in the introduction or reinforcement of sustainable contents, we asked heads of departments to present their perceptions about this issue.

Perceiving the Demands of the Society Regarding the Incorporation of Sustainable Topics

We asked the department heads to evaluate (on a scale from 1 to 7) the interest of different groups in the introduction or reinforcement of sustainable contents in the syllabus of the courses taught by the department. The value

Table 5.1 Societal demands regarding the incorporation of sustainable topics

	Interest in the introduction or reinforcement of sustainable contents	*Importance of the different groups for the department*
Faculty members	3.74	5.60
Steering committees	3.34	4.63
Students	3.47	3.47
Deans	3.86	5.00
Regulators	3.88	4.21
Ecological groups and NGOs	5.17	2.05

"1" indicates "no interest" while the value "7" represents a "very high interest." In Table 5.1, we see that the heads of department perceive that interest in sustainability issues in virtually all of the different groups analyzed is not too high. The expected values are below 4 for the following groups: faculty members, steering committees, students, deans, and regulators. Only ecological groups and NGOs are considered to have a high environmental interest (5.17 out of 7).

In addition, we asked respondents to evaluate (on a scale from 1 to 7) their perceptions about the importance of each group for the decision making in the department. The results are reported in Table 5.1. We can observe that it is precisely the group with the greatest interest in the introduction of sustainable contents, environmentalists and NGOs, that is least important for department managers when making decisions.

Head of Department's Preferences and Sustainable Contents

As mentioned, the heads of departments in Spain are largely responsible for the specific syllabus and orientation of courses, whereas other administrators (e.g., deans) play a broader role, coordinating departments and school management, and serve as symbolic representatives. For this reason, we consider it to be relevant to analyze jointly their preferences and prospects for the future.

In relation to the preferences of the heads of departments, we established three groups on the basis of whether the individuals' preferences for including sustainability-related content in the curricula were above average, below average, or about average (provided their value was within the confidence interval at 95 percent). Table 5.2 shows the current importance of sustainability-related content to the three groups of department heads and their forecasts about the incorporation of sustainable contents in the future. We can see that those heads of departments whose preferences for the

Table 5.2 Head of department's preferences to integrate sustainable contents

	Current sustainable contents			*Future sustainable contents*		
	N	*Average score*	*s.d.*	*N*	*Average score*	*s.d.*
1. Preferences < average	20	1.58	0.526	19	1.85	0.603
2. Preferences = average	10	1.68	0.652	9	2.27	0.816
3. Preferences > average	28	2.07	0.567	27	3.02	0.817
Total	58	1.84	0.619	55	2.49	0.917

Table 5.3 ANOVA test results

	Sum of Squares	*df*	*Mean Square*	*F*	*Sig.*
Between groups	15.694	2	7.847	13.766	.000
Within groups	29.643	52	.570		
Total	45.337	54			

Table 5.4 Scheffé test results

	Nº	*Average score*	*s.d.*	*Scheffé test*
1. Pref < average	19	1.85	0.603	Significant differences between 1 and 3
2. Pref = average	9	2.27	0.816	Significant differences between 2 and 3
3. Pref > average	27	3.02	0.817	Significant differences between 1 and 3 Significant differences between 2 and 3
Total	55	2.49	0.917	

inclusion of sustainability-related content in the curriculum are below the average belong to departments in which the "current sustainable contents" are scarce. Furthermore, their forecasts about the future incorporation of sustainable contents into the curriculum are also very low. At the other extreme, heads of departments whose preferences for integrating sustainable contents are above the average belong to departments in which there currently exists a higher level of sustainable content in the curriculum, and they forecast that even greater inclusion of such material will occur in the future.

An ANOVA analysis was conducted to determine the effect of the department heads' preferences to integrate sustainable contents on the prevision of integration of sustainable contents. The ANOVA test results (Table 5.3 and Table 5.4) show that there were statistically significant means differences among the three groups considered. Thus, those department heads who show a higher personal interest in relation to the incorporation of sustainable contents forecast that, in the next curricula, sustainable contents will also play a broader and more relevant role. The prevision of the department heads in terms of integrating sustainability issues into the courses of the department differ on the basis of their own preferences about integrating sustainability issues into the courses of the department.

Conclusion

At present it appears that the integration of sustainability into normal courses and research projects in universities is the most promising and the most practical way of equipping students with effective tools and knowledge to work for sustainability in their future careers. However, only a small percentage of MBA programs have integrated sustainability features into their courses, and only a few have been explicitly certified as sustainable MBAs. Data from the Aspen Institute show that most of the sustainable MBA programs that it could identify are in the United States, Canada, and the United Kingdom, with a number of Western countries having only one if any such program.

An empirical analysis of the situation in Spanish business schools indicates that environmental knowledge and skills do not receive high levels of treatment in their curricula. Such material appears to receive only residual attention in the current curricula, even when those topics are expected to receive substantially higher attention in the future. Department heads differ in their preferences about incorporating sustainability material into the curriculum and in their predictions about how much such material will be incorporated in the future. Those who prefer more, anticipate more will occur. Those who prefer less, anticipate less will occur.

Although we believe department heads should pay attention to the demands of society for incorporating sustainable topics into the courses taught by the department, we observe that the group of stakeholders, ecological groups, and NGOs, most supportive of increased attention to global sustainability issues are seen by department heads as having the least influence on their decisions.

These results are useful not only for scholars who are analyzing the integration of new topics into business education but also for policy makers and university managers. Understanding the means by which formal and informal support are provided to departments that are integrating sustainability issues is useful for consolidating sustainability progress in education.

References

Appel, G., I. Dankelman, and K. Kuipers. 2004. Disciplinary explorations of sustainable development in higher education. In *Higher education and the challenge of sustainability. Problems, promise, and practice,* ed. B. P. Corcoran and A. E. J. Wals: 213–222. Dordrecht, The Netherlands: Kluwer Academic Publishers.

Aspen Institute. 2009. *The sustainable MBA: The 2010–2011 guide to business schools that are making a difference.* Aspen, CO: Aspen Institute Center for Business Education.

Aznar, P., and M. A. Ullls. 2009. University and Sustainability: Education for Basic Competences. *Journal of Education.* Spanish Ministry of Education: Madrid.

Benn, S., and D. Dunphy. 2009. Action research as an approach to integrating sustainability into MBA programs: An exploratory study. *Journal of Management Education* 33(3): 276–295.

Bradfield, S. L. 2009. The value of sustainability education. *Journal of Management Education* 33(3): 372–375.

Elkington, J. 1997. *Cannibals with forks.* Oxford, UK: Capstone.

Gundlach, M. J., and S. Zivnuska. 2010. An experiential learning approach to teaching social entrepreneurship, triple bottom line, and sustainability: Modifying and extending practical organizational behavior education (PROBE). *American Journal of Business Education* 3(1): 19–28.

Hart, S. L. 2008. Sustainability must be integral to schools' DNA. *Financial Times,* October 13, 2008: 15, London.

Hind, P., A. Wilson, and G. Lenssen. 2009. Developing leaders for sustainable business. *Corporate Governance* 9(1): 7–20.

Holt, D. 2003. The role and impact of the business school curriculum in shaping environmental education at Middlesex University. *International Journal of Sustainability in Higher Education* 4: 324–343.

Juárez-Nájera, M., H. Dieleman, and S. Turpin-Marion. 2006. Sustainability in Mexican higher education: Towards a new academic and professional culture. *Journal of Cleaner Production* 14: 1028–1038.

Lidgren, A., H. Rodhe, and D. Huisingh. 2006. A systemic approach to incorporate sustainability into university course and curricula. *Journal of Cleaner Production* 14: 797–809.

Peoples, R. 2009. Preparing today for a sustainable future. *Journal of Management Education* 33(3): 376–383.

Porter, T., and J. Córdoba. 2009. Three views of system theories and their implications for sustainability education. *Journal of Management Education* 33(3): 323–347.

Rands, G. P. 2009. A principle-attribute matrix for environmentally sustainable management education and its application: The case for change-oriented service-learning projects. *Journal of Management Education* 33(3): 296–322.

Sammalisto, K. 1999. Greening experiences and ambitions as seen by students at some Swedish universities. http://rhu.se/envir/sammalisto_april99.pdf (accessed February 15, 2010).

Sammalisto, K., and K. Arvidsson. 2005. Environmental management in Swedish higher education. Directives, driving forces, hindrances, environmental aspects and environmental co-ordinators in Swedish universities. *International Journal of Sustainability in Higher Education* 6: 18–35.

Sammalisto, K., and T. Lindhqvist. 2008. Integration of sustainability in higher education: A study with international perspective. *Innovative Higher Education* 32: 221–233.

Segalàs, J., Y. Cruz, and K. Mulder. 2004. What professionals should know about sustainable development. ENCOS. European network conference on sustainability practice, Berlin, Germany.

Starkey, R., and R. Welford. 2001. Conclusion. Win–win revisited: A Buddhist perspective. In *Business and sustainable development*, ed. R. Starkey and R. Welford: 353–357. London: Earthscan.

Steketee, D. 2009. A million decisions: Life on the (sustainable business) frontier. *Journal of Management Education* 33(3): 391–401.

Sterling, S. 2004. Higher education, sustainability, and the role of systematic learning. In *Higher education and the challenge of sustainability. Problems, promise, and practice*, ed. B. P. Corcoran and A. E. J. Wals: 49–70. Dordrecht, The Netherlands: Kluwer Academic Publishers.

UN. 2002. The UN Conference on Environment and Development: A guide to Agenda 21, Geneva. UN Environment Programme.

UNEP. 2007. World Environment Report. Nairobi: UNEP.

Walck, C. 2009. Integrating sustainability into management education: A dean's perspective. *Journal of Management Education* 33(3): 384–390.

Walker, H., S. Gough, E. Bakker, L. Knight, and D. McBain. 2009. Greening operations management: An online sustainable procurement course for practitioners. *Journal of Management Education* 33(3): 348–371.

WCED. 1987. *Our common future*. Oxford: Oxford University Press.

Wright, T. S. A. 2004. The evolution of sustainability declarations in higher education. In *Higher education and the challenge of sustainability. Problems, promise, and practice,* ed. B. P. Corcoran and A. E. J. Wals: 7–19. Dordrecht, The Netherlands: Kluwer Academic Publishers.

CHAPTER 6

The Contribution of Interdisciplinary Skills to the Sustainability of Business: When Artists, Engineers, and Managers Work Together to Serve Enterprises

Vera Ivanaj and Silvester Ivanaj

Introduction

Today, with the globalization and internationalization of the world's economy, achieving a globally sustainable present and future appears to many to be the most pressing of all planetary challenges. The principal actors in our society, that is, political, economic, and social actors, are seeking new ways to address the most important challenges of global sustainability: to find viable paths to long-term, socially just development in the world economy, currently in crisis, while preserving our planet from the imminent global warming that could trigger an ecological cataclysm. Social issues are as pressing as economic ones in the pursuit of ways to satisfy the needs of the many actors in society as inexpensively as possible, to preserve and generate jobs, to respect human rights and individual integrity, to fight against discrimination, and to help the developing economies to progress.

In facing these very complex sustainable development (SD) challenges, a consensus seems to have been reached on the need for an integrative approach (Dovers 2005; Nicolescu 2010). This integrative approach concerns both academic disciplines and international political or economic systems, public or private organizations, and various types of interventions. As stressed by Dovers (2005: 3), "The imperative of integration stems from recognition of the interdependence of human and natural systems, expressed in the research and policy agendas of sustainability. International and national policy and law state the 'policy integration principle': environmental, social, and economic considerations must be integrated in decision-making processes to advance the higher-order social goal of an ecologically sustainable, socially desirable, and economically viable future. . . . " In fact, integrating the three pillars of global sustainability or SD—economic, environmental, and social—requires the creation of an *integrative capacity* that allows "a sophisticated understanding of the interactions between highly complex, non-linear, and often closely interdependent human and natural systems" (Dovers 2005: 3).

According to Dovers (2005), integration can find concrete expression at two levels: *informative* and *decisive.* At the informative level, that is, the level of knowledge creation, integration can occur through such actions as the adaptation of a multidisciplinary and interdisciplinary approach in education or the search for global sustainability. At the decisive level, integration can occur in the formulation and implementation of policies and strategies that allow a transition to action and make the principles of global sustainability capable of application. For the first level of integration, educational systems can play a crucial role. As stressed by Clugston (2004: 421), "[S]ustainability challenges universities around the world to rethink their missions and to re-structure their courses, research programs, and the way life on campus is organized. Graduates are increasingly exposed to notions of sustainability that are emotionally, politically, ethically, and scientifically charged. They must be able to deal with conflicting norms and values, uncertain outcomes and futures, and a changing knowledge base. At the same time, they will need to be able to contextualize knowledge in an increasingly globalized society."

Thus, in the context of the "Decade of Education for Sustainable Development" (United Nations Educational, Scientific and Cultural Organization 2004), and under the impetus of recent international and national policies on SD education, many higher educational establishments throughout the world and in many different domains (e.g., engineering science, politics, geography, management, sociology) have become actively involved in the creation of knowledge and skills relating to SD (Wankel and Stoner 2009). A number of them have opted for an inter-, multi-, or trans-disciplinary theoretical approach, allowing them to create and use interdisciplinary skills to resolve problems in SD (Edwards 2009). Most often, these interdisciplinary teams have been created through bringing together engineers, legal specialists, economists, or managers. The fact that artists are members of some teams is often overlooked and little is known about the roles of artists on such teams.

Art and artists can influence the SD of organizations in several different ways, such as the aesthetic appearance and architecture of workspaces, product design, marketing, and advertising. Art and artistic methods can also be important elements in raising awareness, on the part of both individuals and organizations, of the emotional and sensory aspect of SD. Art, as an important trigger of human emotions, can also generate different forms of knowledge and organizational learning useful for SD. As pointed out by Bathurst and Edwards (2009: 119), "[T]o take the radical steps necessary to become sustainable we need an entirely different approach: one that wakes us out of our slumbering complacency and challenges us to transform dramatically the ways in which we conceive of, and practice, business in the twenty-first century . . . it is the artist who offers the social critique and impetus needed to courageously confront realities we would sometimes rather ignore."

The goal of this chapter is to share an innovative educational and institutional approach for combining disciplines in the service of SD: the ARTEM project. ARTEM is a unique experience in which art, technology, and management come together in an educational, research, and institutional project involving three prestigious academic institutions (École Nationale Supérieure d'Art de Nancy, École des Mines de Nancy, and ICN Business School), working with a consortium of about thirty important companies called "ARTEM Entreprises." This project brings together artists, engineers, and managers to create and use transdisciplinary knowledge and competencies as a vector of competitiveness and sustainability for enterprises. We will argue that the junction of these three expertise domains promotes experimentation, creation, and implementation of innovative actions through the emergence of creative and innovative atmospheres and through continuous contact with professional environments.

In this context, our discussion starts by addressing the links that exist between the interdisciplinary approach and SD. The aim is to present the strategic and operational challenges that today's enterprises have to face when first committing to and then implementing an SD strategy. The content of the actions that businesses have to undertake is so complex that only an integrative approach using multidisciplinary, transdisciplinary, and interdisciplinary perspectives and skills will yield a proper understanding of, and ways of addressing, the current SD challenges. Then, we will present the ARTEM-Nancy case. The ARTEM-Nancy project illustrates the benefits of adopting an innovative, integrative, and interdisciplinary approach to education in SD and the role such education can play in assisting enterprises as they develop and implement sustainable strategies.

Global Sustainability and the Interdisciplinary Approach: Essential Dynamic Relationship

The decision-making process for developing an enterprise's SD strategy is highly complex and reflects objectives that vary according to the organization's SD challenges and the strategic level at which these decisions may be taken. Understanding the nature of the key SD challenges organizations face highlights the need to adopt an integrative and interdisciplinary approach within organizations. It also highlights the need for such an approach to creating educational experiences for assisting organizations in developing and implementing effective SD strategies.

The Strategic SD Challenges Facing Enterprises

When talking of global sustainability and SD, the official definition most frequently referred to in the literature is that introduced in the Brundtland Report: "Our Common Future." SD "meets the needs of the present without compromising the ability of future generations to meet their own needs" (World Commission on Environment and Development 1987: 8). This phrasing embodies a change in perception of the paradigm that encompasses the relationship of the economy, economic development, the natural environment, and society. The neoclassical concept of the economy, which has prevailed historically in economic thinking, is one of an economy functioning within a closed system. As pointed out by Stead and Stead (2008: 67), "Depicting the economy as closed implicitly assumes that it is isolated and independent from the social system and ecosystem. Under such an assumption, the economy is not subject to the physical laws of the universe, the natural processes and cycles of the ecosystem, or the values and expectations of society. Assuming these factors away leaves humankind with a mental model of an economy that can grow forever as self-serving insatiable consumers buy more and more stuff from further and further away to satisfy a never ending list of economic desires without any serious social and ecological consequences."

The updated concept of today's economy, at the dawn of the SD age, is one of an economy functioning in an open system, closely linked to nature and society. Starting with that premise, the first definitions of "sustainability" and SD strategies began by taking into account the ecological or environmental dimension. For example, Stead and Stead (1992: 168–169), in their work *Management for a Small Planet,* define sustainability as "a core value [that] supports a strategic vision of firms integrating their need to earn an economic profit with their responsibility to protect the environment." In this context, the winning strategies would be those allowing enterprises to acquire a competitive advantage while at the same time improving the quality of the ecosystem and their own long-term business durability.

The managerial concept of SD has changed greatly, especially following the works of Stuart Hart (Hart 1995, 1997, 2005). Hart helped draw attention in sustainability dialogues to another part of the world and the world economy, all too frequently forgotten: the developing economies. According to Hart, SD strategies are not compiled solely to ensure the success and efficiency of organizations that have decided to develop their activities in developed nations. These strategies must also take account of the economic, social, and environmental concerns of developing nations. The economies of these nations offer new opportunities for development of technology and new products for multinational enterprises whose market share is expanding and extending to new clients—clients with concerns and expectations different from those of the developing world. In this view, remedying basic problems such as famine, epidemics, and inadequate water supplies create new opportunities for business development.

Finally, our concept of SD was widened through the *Triple Bottom Line* approach introduced by Elkington (1997), which provided a global and more integrative vision of SD by combining its three pillars. As pointed out by Stead and Stead (2008: 73), enterprises must thus be capable of introducing "strategic management processes that are economically competitive, socially responsible, and in balance with the cycles of nature. Whereas in traditional strategic management, the term, sustainable, is typically used in reference to a firm's ability to continuously renew itself in order to survive over the long-term . . . we're taking a more comprehensive global view of the term, referring not only to the survival and renewal of the firm itself, but also to the survival and renewal of the greater economic system, social system, and ecosystem in which the firm is embedded."

The social dimension of SD is finding an increasingly prominent position in political thought, becoming an acknowledged strategic objective for businesses as important as economic growth and environmental protection. Businesses are now being required to include in the decision-making process the expectations of other actors—apart from shareholders: consumers, suppliers, employees, associations, financiers, political institutions, et cetera. The entire community is therefore becoming a principal actor that influences the operational activities and strategic decisions of businesses. Concerns such as climate change, child labor, human rights are now entering into the strategic plans of businesses.

Faced with these multiple SD challenges, enterprises have been compelled to draw up and implement SD strategies. The objectives are different depending on the level (corporate, competitive, or functional) and the nature of the strategies to be drawn up. When examined at the corporate level, the aim of the company is to implement strategies that help the business to build a better and sustainable world. According to Stead and Stead (2008: 76), "Corporate level strategies can include creative, generative processes designed to manage the configuration of the firm's portfolio of strategic business units in ways that create synergy among the firm's economic, social and ecological performance." The strategic challenges that this perspective creates for the business are manifold, targeting both "eco-efficiency" and "socio-efficiency." Examples of the wider strategic objectives that can emerge include the preservation of human capital, the stimulation of growth through the development of new markets in the developing world, reducing the waste from excessive consumption in developed countries, savings on natural resources (energy, water, air).

At the competitive level, the aim of SD strategies is to allow the business a durable competitive advantage. Two types of strategies are frequently mentioned in the literature:

1. "*Process-driven*" (pollution prevention), designed to improve environmental efficiency by reducing resource depletion, materials use, energy consumption, emissions, and effluents. Some examples are "redesigning pollution and waste control systems, redesigning production processes to be more environmentally sensitive, using recycled materials from production processes and/or outside sources, and using renewable energy sources" (Stead and Stead 2008: 71).
2. "*Market-driven*" (product stewardship), designed to differentiate products from competitors in the marketplace. Some of the activities are "entering new environmental markets or market segments, introducing new environmentally oriented products, redesigning products to be more environmentally sensitive, advertising the environmental benefits of products, redesigning product packaging, and selling scrap once discarded as wastes" (Stead and Stead 2008: 71).

Finally, at the functional level, the focus in organizations is on balancing the "chain value" between "primary" and "support" activities. An essential tenet of these strategies is Life Cycle Analysis, a total systems approach "designed to provide a cradle-to-grave appraisal of the ecological impacts of a firm's product or service across the firm's value chain" (Stead and Stead 2008: 75). "Among the functional level strategies currently associated with the primary activities of the value chain in strategic environmental management are Design for the Environment (DfE), TQEM, and environmental marketing.

Some of the functional level SSM strategies associated with the support activities of the value chain include full cost accounting, environmental auditing, environmental reporting, Environmental Management Information Systems (EMIS), and human resource management systems" (Stead and Stead 2008: 75).

The Integrative, Interdisciplinary, and Transdisciplinary Approach: A Necessity for Education in SD

Ever since the concept of SD found its way onto the agendas of leading political, economic, and institutional players, the question of how to achieve global sustainability has often been addressed in literature. "Education for Sustainable Development" (ESD) is considered to be an essential tool. In this way, Agenda 21, which was adopted in 1992 in Rio de Janeiro at the United Nations Conference on Environment and Development, insisted on the importance of education in implementing SD principles, notably through entrenching respectful values and attitudes to the environment and to society as a whole. Since 1992, a consensus on the crucial role of ESD has been maintained at all United Nations conferences. The United Nations Economic Commission for Europe defines ESD as follows: "Education for sustainable development develops and strengthens the capacity of individuals, groups, communities, organizations, and countries to make judgments and choices in favor of SD. It can promote a shift in people's mindsets and in so doing enable them to make our world safer, healthier and more prosperous, thereby improving the quality of life. Education for SD can provide critical reflection and greater awareness and empowerment so that new visions and concepts can be explored and new methods and tools developed" (United Nations Economic Commission for Europe 2005, p. 1). The Johannesburg Summit (World Summit on Sustainable Development 2002) reaffirmed the objectives of ESD by proposing the idea of a "Decade of Education for Sustainable Development." At the UN General Assembly in December 2002, this concept was defined as having the overall aim of integrating SD principles, values, and practices into all areas of education and apprenticeship (United Nations Educational, Scientific and Cultural Organization 2004).

What is encountered most frequently in the literature on education and research for SD is the need to adopt an integrative, interdisciplinary, or transdisciplinary approach (McNeill 2001; Eishof 2003; Marinova and McGrath 2004; Dovers 2005; Polk and Knutsson 2008; Edwards 2009; Rohwer 2010; Nicolescu 2010). For example, Nicolescu (2010: 1) postulates: "If the universities intend to be valid actors in SD they have first to recognize the emergence of a new type of knowledge—the transdisciplinary knowledge—complementary to the traditional, disciplinary knowledge. This process implies a necessary multi-dimensional opening of the University: towards the civil society; towards the other places of production of the new knowledge; towards the cyber-space-time; towards the aim of universality; towards a redefinition of values governing its own existence." The concepts of "transdisciplinarity" and "interdisciplinarity" have emerged in the sciences because of the need to build bridges between the various disciplines. As Nicolescu (2010: 1) indicates, these concepts differ from those of "multidisciplinarity" or "pluridisciplinarity": "Pluridisciplinarity concerns studying a research topic not in only one discipline but in several at the same time." "Interdisciplinarity has a different goal from multidisciplinarity. It concerns the transfer of methods from one discipline to another." Finally, "transdisciplinarity concerns that which is at once between the disciplines, across the different disciplines, and beyond all discipline. Its goal is the understanding of the present world, of which one of the imperatives is the unity of knowledge."

The need to adopt a multidisciplinary or transdisciplinary approach is closely linked to the very nature of SD, which requires the simultaneous understanding and integration of three different and complex dimensions that interact with each other: economic, ecological, and social. These three dimensions give rise to different objectives (Serageldin and Steer 1994: p. 1): (1) economic objectives (growth, equity, and efficiency); (2) ecological objectives (ecosystem integrity, carrying capacity, biodiversity, and global issues); and (3) social objectives (empowerment, participation, social mobility, social cohesion, and institutional development). In addition, each of these three dimensions can be connected to distinct scientific disciplines with their contrasting theoretical and methodological perspectives: economics, anthropology (or sociology), and ecology (McNeill 2001: 3). For example, "Economics is concerned with human beings interacting with each other as (rational, self-interested, autonomous, maximizing) decision-makers, with the emphasis on the individual entity; nature is typically treated as a material resource/constraint. Anthropology is concerned with human beings interacting with each other not only as decision-makers but also as meaning-makers, with the emphasis on the collective; nature is regarded both as a resource/constraint and as a locus of meaning. Ecology is concerned with human beings as a species, interacting as biological beings, both with their own and other species and with the inorganic environment; the emphasis is on the whole as a system" (McNeill 2001: 3). In addition, according to McNeill (1999), these three disciplines can interact to produce hybrid disciplines or subdisciplines, such as environmental economics, ecological economics, environmental anthropology, sociological economics, sociobiology. These subdisciplines are also effective in bringing new perspectives and different solutions to the problem of SD, and are of great interest when cultivating a global and integrative understanding of the problem.

In this context, interdisciplinarity and transdisciplinarity have become the natural choice in education for SD in universities. As pointed out by Marinova and McGrath (2004: 2), this perspective "increases awareness of the complexity and interrelationships of environmental, economic, social, political, and technical systems and also increases respect for the diversity of voice that exists amongst cultures, race, religion, ethnic groups, geographic, and intergenerational populations. . . . This complexity and diversity in the world requires knowledge and skills by citizens, professionals, and leaders that cross the boundaries of disciplines and institutions, cultures and realities of society."

In this context of recognized usefulness of interdisciplinary and transdisciplinary approaches in education and research for SD, a number of researchers have already documented in their works experiments to develop interdisciplinary or transdisciplinary training programs. For example, the work of Rudebjer (2008) describes the experience and challenges of designing and implementing a program of education for natural resource management and sustainable agriculture at the Agricultural University of Norway. In this program, interdisciplinarity was a principal ambition. Flint, McCarter, and Bonniwell (2000) describe the Northampton Environmental Legacy Program, which links studies of the historic culture of Eastern Shore life with an awareness and understanding for the importance of environmental quality in that region. Similarly, Posch and Steiner (2006) use a transdisciplinary case study conducted in Austria to describe and illustrate the task of integrating research and teaching on sustainability-related innovation. Tomkinson, Tomkinson, Dobson, and Engel (2008) describe an interdisciplinary pilot program in SD for University of Manchester undergraduate engineers and scientists. Costa and Scoble (2006) report how the Mining Engineering Department at the University of British Columbia is addressing the need to integrate SD into mining engineering on behalf of industry and society. Their study describes the evolution of an interdisciplinary model, the Sustainability Working Group, which brings together a diverse array of disciplines from academia, industry, government, NGOs, and mining communities.

The ARTEM-Nancy Case: Integrating "Art, Technology and Management" for Global Sustainability

The ARTEM-Nancy experience is in the tradition of those other innovative approaches to education for global sustainability. The ARTEM-Nancy case illustrates the way three university partners have joined with other institutions in an interdisciplinary alliance to create an innovative educational project that gives students key skills in the field of SD.

ARTEM-Nancy: History, Origin, Philosophy, and Education Policy

ARTEM is an innovative interdisciplinary, intercultural, scientific, and institutional alliance established between three major educational institutions in Nancy, France: École Nationale Supérieure d'Art de Nancy, École des Mines de Nancy, and ICN Business School. The alliance was started in 1999 by three Nancy universities that welcome some 45,000 students each year: the National Polytechnic Institute of Lorraine, the University of Letters and Human Science Nancy 2, and the Scientific and Medical University UHP Nancy 1. The alliance is supported by the public regional authorities concerned: the Urban Community of Grand Nancy, the General Council of Meurthe-et-Moselle Department, and the Regional Council of Lorraine. The ministers responsible for culture, national education, and industrial research have also lent their support through written undertakings. In addition, ARTEM enjoys the support of a significant number of businesses (currently 31) and other local and national economic actors who have come together in this context in an association known as ARTEM-Entreprise (ARTEM-Nancy 2010).

ARTEM is first and foremost an alliance of Art, Technology, and Management much inspired by the local cultural heritage of the Lorraine Region: Art Nouveau, which marked the blending of art and commerce in the late nineteenth century, and the École de Nancy, which founded the Alliance Provinciale des Industries d'Art. ARTEM is designed to meet more effectively the economic challenges of the twenty-first century, a time when businesses are seeking ever more complex goods and services for a globalized and internationalized economy with a future that is becoming steadily less predictable. Innovation in both technology and products is becoming a key success factor in this context, and the creative capacities of businesses are now a differentiating factor (Cremet, 2006). ARTEM is also responding to the development needs of the regional economy in which competition between economic actors is intense. In this environment, great importance is attached to nonmaterial technology and to knowledge for developing new products and services. The development and use of creative skills in the service of design and creative industries is thus of paramount importance. ARTEM responds directly to the current needs of firms looking to recruit managers, engineers, artists, or designers capable of working in the multicultural and multidisciplinary teams required for achieving creative solutions where disciplines intersect (Cremet 2006). Although the alliance between engineers and managers is more widespread, the originality of the ARTEM project compared to others of this type lies in the fact that artists play a key role in this project. Artists are now called upon to face the emergence of new technologies, new materials, and new perceptions of their jobs. In their emerging jobs, interactivity with individuals and communication with the outside world is playing a role of ever-increasing importance. The potential for art and aesthetics to contribute to a sustainable world is also starting to be recognized.

In this historical, economic, and cultural context, "ARTEM is an original initiative that combines the creation and integration of new technologies and managerial, strategic, economic, and legal perspectives" (ARTEM-Nancy 2010). Its basic philosophy is: "To unite creators, engineers and managers is to prioritize the dialogue of cultures, it is to make the knowledge system a real factor in competitiveness, it is to build innovation with an ethic that combines synergy and education, strategy and empathy" (ARTEM-Nancy Website 2010). As emphasized by Claude Cremet, former dean of the École des Mines de Nancy and one of the founders of this alliance, ARTEM is synonymous with educational innovation in terms of pluridisciplinarity and interdisciplinarity. This interdisciplinarity is reflected especially in the introduction of "ARTEM workshops," which represent "educational platforms for interdisciplinary and intercultural meetings between students and lecturers within schools in contact with the business environment" (Pléty and Cremet 2008: p. 2). Since 2000, 20 Research and Creation workshops have been set up between the three schools. These workshops are open to all students from the three schools. About 550 students are eligible to attend the workshops each academic year: the 350 students in the penultimate year of the Master's course at the ICN Business School, the 150 students in the École des Mines de Nancy, and the 50 students in their last two years of training at the École Nationale Supérieure d'Art de Nancy.

The workshops are held throughout the academic year (September to May) on a block day in the week. Each day is divided into two sessions: an initial conference session led by a university or representative from the business world is followed by a joint project teamwork session. Each workshop offers different themes that develop over time with the stress on developing interdisciplinary skills.

The ARTEM workshops are also open to businesses that are invited to introduce people and problems from specific projects, issues of concern to the businesses, and research or business development issues they are facing. Some workshop projects have led to the creation of businesses, and some have received awards in national and regional business development competitions. Each workshop offers 150–180 effective working hours. In the

20 workshops, some 3,000 hours of teaching have been provided by the three schools (ARTEM-Nancy 2010). Each year, about 400 different people are involved in 80 projects in cross-cutting teams.

In addition to the workshops, the three schools organize an annual weeklong business simulation game. During this game, 40 interdisciplinary teams compete virtually on the world market by creating a new product. Associative projects common to all three schools are also offered, such as participation in "*Défi Voile Artem*," the national sailing competition in which students from schools throughout France come together in a sporting and managerial contest under extreme conditions.

Another feature of the ARTEM-Nancy alliance is the creation, in 2000, of the "ARTEM-Entrerprises" association. This 1901 Law association currently includes 31 local, national, and international businesses representing such sectors as energy (e.g., EDF, Veolia), finance (e.g., CIC, BNP Paribas), office equipment (Xerox), large-scale distribution (Cora). This association supports the ARTEM dynamic and provides the forum for dialogue between students, teachers, researchers, and professionals. These dialogues help schools design and improve the content of their training schemes by bringing together the needs of businesses and the most recent developments in the world of work and international markets. The alliance businesses are also potential and effective sources of employment for students trained within the ARTEM framework (ARTEM-Nancy 2010).

In addition to the opportunity for daily dialogue and collaboration, ARTEM-Entreprise also offers facilities for developing cross-cutting, interdisciplinary, and transdisciplinary skills in "cross-cutting workshops." Starting in 2007, several businesses have been offering both students and teachers from the three schools problems or concerns as themes for cross-cutting workshops. These workshops allow students to offer specific solutions to these business problems while working together in a group, with professionals and lecturers sharing their experience and expertise. Two recent workshops are particularly worthy of mention: "De-materialization of the workspace," a workshop organized in partnership between Gan Assurances, Société Générale, and Vosges Matin, and "Building a 200-person Headquarters," a workshop offered in partnership between Pertuy, GDF SUEZ, and Veolia Transport (ARTEM-Nancy 2010).

The Contribution of ARTEM-Nancy to the Acquisition of Interdisciplinary Skills in the Service of SD

A close look at the subjects of the ARTEM workshops in Table 6.1 reveals that its workshops help students acquire interdisciplinary skills in the three

Table 6.1 ARTEM workshops offered by the three schools

School of Mines
• Environment and SD
• E-business
• Regional engineering and social innovation: environment and actors
• Business and regions
• Cindynics or the science of danger
• Design, innovation, production: "from idea to prototype"
School of Fine Arts
• Shared works (season 4)
• The material and nonmaterial landscape: devices for looking at the *Colline de Sion*
• Cohabitation electroshop: interactive devices
• Body present, body absent (embraces/stranglehold)
• Direct and peaceful action: discovering and practicing "design activism"
ICN Business School
• CFA workshop
• A "Business Innovation" Chair
• Organization of large-scale distribution and the mass consumption products industry
• Controlled management of technological innovation
• Economic intelligence and decision making
• "Work Space 2.0"
• Financial modeling

Data compiled by theauthors from information on the ARTEM-Nancy Website (ARTEM-Nancy Website, 2010)

essential aspects of SD: the economic, environmental, and social dimensions. These skills are closely linked to the piloting and implementation of SD strategies.

The workshops offered by the École des Mines de Nancy principally focus on the management of the environmental or ecological aspect of SD with a theoretical and pragmatic approach. The SD of urban or rural areas, transportation, sustainable habitat, and the management of major industrial hazards are also covered in depth in these workshops. The content of the "environment and sustainable development" workshop provides an excellent illustration of the importance given to the environmental aspect of SD. The context of two of the very comprehensive modules is shown in Table 6.2.

The two modules in Table 6.2 stress the principal restrictions and grounds that compel businesses to implement SD strategies, such as regulatory restrictions at national or European levels connected with protecting the environment, saving energy, reducing pollution and greenhouse gases, and reducing solid and liquid waste.

Table 6.2 Content of the "Environment and Sustainable Development" workshop

Module 1

Introduction to concepts of environmental law through industrial cases
- Hierarchy of legislative texts; the Environment Code; international and European environmental legislation; environmental policy of the EU (e.g., REACH et cetera)
- Installations classified for environmental protection purposes (ICPE) through a case study (visit to an industrial site, examination of the particular case, meeting with actors, etc.)

Introduction to the concept of risk and its acceptability—the concept of risk and toxicity—risk analysis methods—pollutant dispersion

Environmental management system and integrated system
- ISO 14000 standards, Eco-Audit regulation (by industrial case studies)
- Health and Safety Integrated Systems (OHSAS 18001), quality and environment

Sustainable Development
- Introduction to the concept of SD (through role-play games), general ideas, SD 21000 guides, examples (fair business and energy, The Enercoop case, et cetera)
- Environmental performance evaluation methods: life cycle analysis (LCA), use of GaBi 4 Software, and carbon balance sheet method

Particular procedures
- Best available technologies (BAT)
- The greenhouse gas problem: reducing CO_2 emissions, capturing and holding CO_2, Kyoto protocol, quota directive, PNAQ (national quota plan)

Module 2

Introduction: industrial obligations, "end-of-pipe" treatment processes

Treatment processes and liquid effluent control
- Water management and quality in France, water resources, role of water agencies
- The challenges in industrial and urban wastewater and sludge recycling
- Description of treatment processes—choice of section of industry—visit to wastewater treatment station

Treatment processes and solid waste control
- The concept of waste
- Waste as a source of energy—visit to processing units
- Recycling materials (recycling of aluminum, steel et cetera)
- Toxic waste and end residue—stabilization and inertia—visit to RESOLEST factory

Treatment processes and gaseous effluent control
- Atmospheric pollution
- Climate change and the greenhouse effect
- Gaseous effluent treatment processes—self contained treatment operations

Energy and environment
- Energy industry, situation concerning renewable energy in France and in the world
- Energy and CO_2: the renewable energy solution (solar, wind, sea energy et cetera)—visit to a wind farm and biomass recycling
- Hydrogen as an energy source
- Problem: heating the living space
- Nuclear energy and the problem of radioactive waste (visit to waste storage center)

Data compiled by the authors from information on the ARTEM-Nancy Website (ARTEM-Nancy Website, 2010)

The workshops offered by the École Nationale Supérieure d'Art de Nancy are of particular interest in understanding the social dimensions of SD. Art and aesthetics help us understand the action of humans on nature, on their own personal and interpersonal spaces, and on other actors in society as a whole. The study of artistic or creative works is used as a means of conveying a sensory message capable of influencing the behavior of individuals in society. Such study also allows a better emotional understanding of acts of influence, discrimination, persecution, subjugation, et cetera. through the study of artistic personages and role-playing. In these ways, the workshops help students reflect on the effect of history and culture on questions of SD. For example, the "shared works" workshop relates to the study of multiuser systems present, in particular, in video games (e.g., *Second Life, World of Warcraft,* and *Solipsis*). It raises the question of "the presence of the other" and of "the idea of collaboration" as elements for creating links and managing the space between individuals. The "material and nonmaterial landscape" workshop helps students understand space in its material dimension (territory and its geophysical components) and nonmaterial dimension (cultural and imaginary aspects, and the imagery of places). The workshop helps provide a knowing and inventive insight into places to assess their possibilities for development through historical, cultural, and scientific research work. The questions raised relate to ways in which a place can be developed sustainably without robbing it of its identity.

Finally, the workshops offered by ICN Business School contribute more to acquisition of skills linked to the economic aspects of SD: how to increase business activity in a global economy becoming more and more internationalized, how to innovate in terms of technologies and products, how to create one's own sustainable-development-oriented business, how to invest in ethical funds, how to dematerialize work spaces, etc. Table 6.3 presents the content of the "Anglo-Saxon Business Projects" workshop. Internationalization and foreign market knowledge are important aspects of this workshop.

As noted earlier, cross-disciplinary SD skills can also be learned via the ARTEM-Entreprises interdisciplinary workshops. The "Dematerialization of workspaces" workshop is one particularly good example of this opportunity. This workshop focuses on the way in which new technologies are incorporated into the work space and the environmental, social, economic, and cultural challenges the work space presents for companies.

This workshop enables students to see that the dematerialization of work spaces has an impact on the business activity itself by decompartmentalizing it and allowing communication and activity to flourish between designers, engineers, managers, and sales personnel. Economic and social benefits can only be achieved if these technologies make sense in the organization of

Table 6.3 Content of the "Anglo-Saxon Business Projects" workshop

Teaching objectives

- Tutor students in a professional project involving the introduction onto an Anglo-Saxon market of a new product or service.
- Introduce students to the business environment in countries where the Anglo-Saxon influence is, or historically has been, strong. The business environment will include aspects of the macroeconomy as they affect business operations and performance; aspects of the microenvironment that will be useful for entrepreneurs and other members of the business community involved in business with companies based in the target country. These will include market characteristics and customer behavior; practical issues concerning insertion into the target business community.
- Introduce students to aspects of business culture in the target country. Business culture will mean looking at what the components of business culture are and how these are developed and communicated to the corporate or organizational stakeholders.
- Introduce students to the theories, models, and practices pertaining to conducting business outside of France. Topics covered will include project management, enterprise creation, and international marketing strategy.

Workshop Content

- Business culture: attitudes, behavior, values;
- Economic and business environment in UK, United States, Canada, India, and Ireland;
- Project management;
- International marketing strategy;
- Creative thinking in business.

In each of the countries studied, discussion will include aspects of the market, entrepreneurship, and consumer behavior.

Student Project

Students will be required, via the project, to develop relations with local, regional, national or international players in the business environment. They will be required to identify, create, or imagine a product or service that they feel will be appropriate for one of the Anglo-Saxon markets they have studied. Examples of such products may include original items of art, design, or manufacture.

Data compiled by the authors from information on the ARTEM-Nancy Website (ARTEM-Nancy Website, 2010)

enterprises and in the environments of those who use them. The successful introduction of these technologies is also conditioned by the social acceptability and technical usability of the new work processes (ARTEM-Nancy 2010). The educational philosophy used in this workshop takes a cross-disciplinary approach involving lecturers from the three schools and seminars with professionals, plus student projects guided by managers from partner companies.

Conclusions

This chapter has considered the dynamic link that exists between interdisciplinarity and SD, and the need to create interdisciplinary skills as an important factor in the design and implementation of global SD strategies. It has also stressed the fact that global sustainability cannot be achieved unless its three core dimensions are integrated and carefully managed, specifically the economic, environmental, and social dimensions. These three dimensions are different yet complementary. Much of the difference stems from the sheer diversity of the basic theoretical and empirical foundations upon which they are conceptualized and enacted, such as ecology, economics, anthropology, and sociology. Segmented or purely multidisciplinary approaches are not always helpful for SD initiatives because they do not reveal the dynamics created between the disciplines and the potential these dynamics offer for solving the complex problems that are inherent to the very nature of the SD phenomenon.

This chapter has also stressed the important contribution educational establishments can play in the inter- and trans-disciplinary knowledge formation and skills development needed for SD. The ARTEM-Nancy project illustrates the value of an educational philosophy that brings together engineers, artists, and managers to work on training initiatives in SD. The project is currently the only one of its kind, but the learning process it has adopted may be a useful one for other institutions to consider. The project creates independent learning through running projects in interdisciplinary groups, supported by professionals from companies and lecturers who bring their own very different contributions. These contributions are simultaneously artistic and creative, managerial and organizational, and technological or engineering. This process enables students to become instrumental in their own training through their choice of complex projects offered in a variety of workshops.

References

ARTEM-Nancy. 2010. http://artem.inpl-nancy.fr (accessed March 21, 2010).

Bathurst, R., and M. Edwards. 2009. Developing a sustainability consciousness through engagement with art. In *Management education for global sustainability* (pp. 115–137), ed. C. Wankel and J. A. F. Stoner. Charlotte, NC: Information Age Publishing, Inc.

Clugston, R. M. 2004. Foreword. In *Higher education and the challenge of sustainability: Problematics, promise, and practice* (pp. ix–xii). Boston: Dordrecht.

Costa, S., and M. Scoble. 2006. An interdisciplinary approach to integrating sustainability into mining engineering education and research. *Journal of Cleaner Production* 14 (3–4): 366–373.

Cremet, C. 2006. L'invetion d'ARTEM-Nancy: récit d'une expérience. http://www.mcxapc.org/docs/cerisy/a4-1.pdf (accessed March 21, 2010).

Dovers, S. 2005. Clarifying the imperative of integration research for sustainable environmental management. *Journal of Research Practice* 1 (2), Article M1, 2005,

Published online by ICAAP. http://jrp.icaap.org/index.php/jrp/article/view/11/30 (accessed March 19, 2010).

Edwards, M. G. 2009. Visions of sustainability: An integrative metatheory for management education. In *Management education for global sustainability* (pp. 51–91), ed. C. Wankel and J. A. F. Stoner. Charlotte, NC: Information Age Publishing, Inc.

Eishof, L. 2003. Technological education, interdisciplinarity, and the journey toward sustainable development: Nurturing new communities of practice. *Canadian Journal of Science, Mathematics and Technology Education* 3 (2): 165–184.

Elkington, J. 1997. Cannibals with forks. Oxford, UK: Capstone.

Flint, W. R., W. McCarter, and T. Bonniwell. 2000. Interdisciplinary education in sustainability: Links in secondary and higher education; The Northampton Legacy Program. *International Journal of Sustainability in Higher Education* 1 (2): 191–202.

Hart, S. 1995. A natural resource-based view of the firm. *Academy of Management Review* 20 (4): 966–1014.

———. 1997. Beyond greening: Strategies for a sustainable world. *Harvard Business Review* 75 (1): 66–76.

———. 2005. *Capitalism at the crossroads: The unlimited business opportunities in solving the world's most difficult problems.* Upper Saddle River, NJ: Wharton School Publishing.

Marinova, D., and N. McGrath. 2004. A transdisciplinary approach to teaching and learning sustainability: A pedagogy for life. Teaching and Learning Forum 2004 (Proceedings contents). http://lsn.curtin.edu.au/tlf/tlf2004/marinova.html (accessed March 20, 2010).

McNeill, D. 1999. On interdisciplinary research: With particular reference to the field of environment and development. *Higher Education Quaterly* 53(4): 312–332.

———. 2001. Inter-disciplinarity and sustainable development policy: What have we learned?. Presentation at the World Bank, 3 December 2001. http://www.sum.uio.no/pdf/publications/working-and-occasional-papers/mcneill.pdf (accessed March 20, 2010)

Nicolescu, B. 2010. The transdisciplinary evolution of the university condition for sustainable development. http://basarab.nicolescu.perso.sfr.fr/ciret/bulletin/b12/b12c8.htm (accessed March 19, 2010).

Pléty, R., and C. Cremet. 2008. Pour une pédagogie de l'innovation. Conference ARTEM, 15 May 2008, Conseil Général de Meurthe et Moselle.

Polk, M., and P. Knutsson. 2008. Participation, value rationality and mutual learning in transdisciplinary knowledge production for sustainable development. *Environmental Education Research* 14 (6): 643–653.

Posch, A., and G. Steiner, 2006. Integrating research and teaching on innovation for sustainable development. *International Journal of Sustainability in Higher Education* 7 (3): 276–292.

Rohwer, G. 2010. How to encourage education for sustainable development in geography. http://www.geo.fu-berlin.de/geog/fachrichtungen/schulgeog/medien/download/RohwerHow_to_encourage.pdf (accessed March 20, 2010).

Rudebjer, P. 2008. Crafting interdisciplinarity in teaching natural resource management and sustainable agriculture. http://www.betuco.be/voorlichting/Enabling%20Participatory%20 Research%20and%20Development.pdf#page=41 (accessed March 21, 2010).

Serageldin, I., and A. Steer. 1994. *Making development sustainable: From concepts to action*. Washington, D.C.: World Bank.

Stead, J. G., and W. E. Stead. 2008. Sustainable strategic management: An evolutionary perspective. *International Journal of Sustainable Strategic Management* 1 (1): 62–81.

Stead, W. E., and J. G. Stead. 1992. *Management for a small planet: Strategic decision making and the environment.* Newbury Park, CA: Sage Publications.

Tomkinson, B., R. Tomkinson, H. Dobson, and C. Engel. 2008. Education for sustainable development—an interdisciplinary pilot module for undergraduate engineers and Scientists. *International Journal of Sustainable Engineering* 1 (1): 69–76.

United Nations Economic Commission for Europe. 2005. UNECE Strategy for Education for Sustainable Development. http://www.unece.org/env/documents/2005/cep/ac.13/cep.ac.13.2005.3.rev.1.e.pdf. (accessed March 20, 2010).

United Nations Educational, Scientific and Cultural Organization. 2004. *United Nations Decade of Education for Sustainable Development 2005–2014.* Draft International Implementation Scheme. Paris: UNESCO.

Wankel, C., and J. A. F. Stoner. 2009. *Management education for global sustainability.* Charlotte, NC: Information Age Publishing, Inc.

World Commission on Environment and Development. 1987. *Our common future.* Oxford, UK: Oxford University Press.

World Summit on Sustainable Development. 2002. *Plan of implementation of the World Summit on Sustainable Development.* New York: United Nations.

CHAPTER 7

Public Disclosure of Corporate Environmental Performance: Pollutant Release and Transfer Registers (PRTRs)

Javier Delgado-Ceballos and Antonio Rueda-Manzanares

Introduction

Global sustainability—"meeting this generation's needs while enhancing the ability of future generations to meet theirs"—is becoming increasingly important in organization theory and practice. Literature has shown that stakeholders and regulatory actions can have significant influence on corporate environmental behavior. Much research has been carried out on the role of stakeholders—such as customers, mass media, and governments, among others—in environmental management and the effectiveness of their interventions (Cordano, Frieze, and Ellis 2004; Sharma and Henriques 2005). Although these and other stakeholders play a critical role in promoting global sustainability, studies show that regulatory stakeholders seem to have the greatest influence on corporate environmental behavior (Buysse and Verbeke 2003; Dasgupta, Hettige, and Wheeler 2000; Henriques and Sadorsky 1996).

Having used traditional command-and-control and market-based approaches to influence corporate environmental behavior, governments now seem to be particularly interested in increasing the use of public disclosure of corporate environmental performance through Pollutant Release and Transfer Registers (PRTRs) (Kerret and Gray 2007). Research studies (Cohen 2002; Tietenberg and Wheeler 2001) indicate that the disclosure of installations' environmental emissions often pressure companies into improving their environmental performance. This approach is based on the belief that once different stakeholders have information about an installation's emissions, they will pressure firms to decrease those emissions (Weil et al. 2006). Bansal and Clelland (2004) show that the information published on PRTRs has an enduring impact on companies: firms that are perceived as environmentally illegitimate will experience an increase in unsystematic risk relative to those seen as legitimate. The PRTRs are also seen as providing information of much greater value in comparative analyses than environmental information published by the companies themselves (Sullivan and Gouldson 2007).

This chapter analyzes how governments pressure firms to improve their environmental performance through collection and disclosure of the information contained in PRTRs. Specifically, the chapter seeks a better understanding of the different alternatives for using the information included in PRTRs to improve corporate environmental performance and to reduce information asymmetries. In doing so, the chapter analyzes how environmental information is presented and how it sends different types of signals about a firm's environmental performance.

Background: Stakeholders' Roles in Environmental Behavior

In his seminal work *Strategic Management: A Stakeholder Approach*, Freeman (1984) defined the concept of stakeholder to include any individual or group who can affect an organization's performance or who is affected by the organization's pursuit of its objectives. This definition includes practically any social agent, individual or collective, related to an organization.

In the past decade, much stakeholder management theory was focused on which stakeholders matter most. The answer seems to lie with those who possess power, legitimacy, and urgency (Mitchel, Agle, and Wood 1997). Manager's perceptions are crucial because those perceptions determine the relative prominence of stakeholders in influencing managerial decisions (Agle, Mitchell, and Sonnenfeld 1999). Stakeholders can be classified in different categories, for example, as primary or secondary stakeholders; or in different groups, for example, community stakeholders, regulatory stakeholders, organizational stakeholders, and the media (Berman et al. 1999; Clarkson 1995; Freeman 1984; Henriques and Sadorsky 1999). As noted above, empirical studies show that pressure from regulatory stakeholders (such as legislative and governmental) appears to play the most influential role in a variety of industries and countries.

Regulatory Stakeholders: Legislative and Governmental

Governmental regulations related to companies' operations have been shown to create pressures that lead those companies to modify their environmental performance (Delmas and Toffel 2004), including such behaviors as pollution prevention strategies (Buysse and Verbeke 2003) and strategic approaches toward the natural environment (Hart 1995; Shrivastava 1995). Regulations have also been shown to influence competitive positions (Porter and van der Linde 1995), financial performance (Hillman and Hitt 1999; Shaffer 1995; Williamson 1979), and internal management practices (Delmas and Toffel 2004). There is a high degree of interdependence between a firm's competitive environment and public policy because regulators can also alter the size or structure of markets and influence product demand through taxes (Baron 1995).

Regulatory Stakeholder: Carrots and Sticks

Governments and legislatures employ multiple "carrots and sticks" to pressure firms to improve their environmental performance (Rugman and Verbeke 1998) and bring about changes in the environmental practices of companies (Dasgupta, Hettige, and Wheeler 2000; Henriques and Sadorsky 1996).

First, researchers have focused on the influence of enforced legislation and regulations on a firm's environmental practices (Majumdar and Marcus 2001; Rugman and Verbeke 1998). Actually, environmental regulations have significantly increased over the past 30 years (Hart 1995; Rugman and Verbeke 1998). These regulations have been moving from command-and-control regulation—based on pollution limits and technological standards—to more flexible positions. Tietenberg (1998) concluded that these methods were costly and unable to fulfill their established aims.

Second, governments have the power to exert criminal enforcement. Breach of law may lead to important penalties, products taken off store shelves, and even firm closure. Kassinis and Vafeas (2002) showed that both direct and indirect costs, provoked by environmental regulation, are so high that they can indeed increase the strategic importance of environmental breaches by firms. In addition, firms incurring environmental penalties are closely followed by governments and environmental groups due to loss of legitimacy. Banned firms will frequently take extra care not to commit further environmental infraction.

Finally, governments have started to apply other methods to affect corporate environmental policies, such as voluntary programs (e.g., U.S. Industrial

Toxics Emission Reduction Program 33/50), market incentives (e.g., U.S. Deposit-Refund System, Pollution Charges Marketable Permit Systems), business awards (EU—the European Business Awards for the Environment), and forced disclosure of environmental information (e.g., PRTRs). This chapter focuses on this last method.

Delimitation of PRTRs

Environmental information disclosure strategies consist of "public and/or private attempts to increase the availably of information on pollution to workers, consumers, shareholders, and the public at large" (Tietenberg and Wheeler 2001). Policies on the public disclosure of corporate environmental performance are quasi-regulatory instruments, or the "third wave" of environmental regulation (Cohen 2002; Tietenberg and Wheeler 2001). In this type of situation, the regulatory environment forces companies to report their chemical emissions and other releases through PRTRs. The Organization for Economic Cooperation and Development (OECD) defined PRTR as a catalog or register of releases and transfers of potentially harmful substances to the environment from a variety of sources (OECD 1996).

Researchers (Cohen 2002; Tietenberg and Wheeler 2001) state that the disclosure of environmental emissions pressures companies into improving their environmental performance. The idea behind this policy is that when shareholders, consumers, and environmental groups have information about an installation's emissions, they will exert pressure to decrease the emissions (Weil et al. 2006). In other words, governments put pressure on those firms showing unacceptable environmental performance by making their emissions public through rankings. In fact, initial evidence shows that disclosure of corporate emissions has reduced emissions in a number of countries (Konar and Cohen 1997; Foulon, Lanoie, and Laplante 2000).

In this context, regulators play a facilitator role rather than a coercive role (Scholz and Gray 1997). This option is perceived as a low-cost regulatory procedure since formal enforcement procedures are not demanded (Antweiler and Harrison 2007; Tietenberg and Wheeler 2001).

PRTR Background: Right to Know and Environmental Information Asymmetry

PRTRs promote the public right to know, monitoring of environmental policy, and the encouragement of reductions in emissions and risk (OECD 2000). PRTRs require industrial installations to measure and report their emissions into water and air and onto land (Sullivan and Gouldson 2007).

Before the enactment of such legislation, corporate environmental data (such as an installation's chemical emissions) were not disclosed publicly. Pressures from various stakeholders, however, forced greater data availability and disclosure through public policies (Gerde and Logsdon 2001). These pressures were mainly based on (1) the demand to know what was happening within industrial plants and (2) the existence of environmental information asymmetry.

Following the disastrous toxic gas leak in Bhopal, India, in December 1984 (Hoffman 2001), communities and environmental activists demanded their right to know what was taking place within industrial plants. Pressures were based on the fact that existing information was incomplete. As a result, workers, consumers, and the public were only partially aware of a given installation's environmental behavior (Henriques and Sadorsky 1996) and their health hazards consequences (Akerlof 1970). Henriques and Sadorsky (1996) stated: "Consequently, they will be unable to trade off higher risks for either higher wages or lower prices in an informed way so that the unaided market would not necessarily result in either the right amount or the correct distribution of risk." It was recognized that making information on plant operations and emissions available to the public could have direct bearing on how facilities conduct business and how they manage their emissions (OECD 2000; OECD 2005).

Legislation resulting from these pressures was based both on human rights and on freedom of access to information. The legislation led governments to develop PRTRs over the past two decades (Sullivan and Gouldson 2007).

A company's environmental information may be asymmetrically distributed between the firm and the community (Kullkarni 2000). Information asymmetry exists when firms have more information about their environmental practices than the community. Therefore, workers, consumers, and community are only partially aware of an installation's environmental behavior (Henriques and Sadorsky 1999), the health hazards (Henriques and Sadorsky 1999), and the impact of a firm's production processes (Kullkarni 2000). The information asymmetry between a firm and the community may be further reinforced by a firm's desire to act opportunistically (Kulkarni 2000).

PRTRs around the World

Several countries have already developed their own PRTRs. Table 7.1 shows the number of PRTRs around the world, their names, and web pages for accessing them.

Table 7.1 List of PRTRs around the world

Country	*Name*	*Web Page*
Australia	National Pollutant Inventory	http://www.npi.gov.au/
Canada	National Pollutant Release Inventory	http://www.ec.gc.ca/inrp-npri/
Chile	Registro de emisiones y transferencia de contaminantes	http://www.conama.cl/retc/1279/channel.html
Europe	European PRTR	http://prtr.ec.europa.eu/
Japan	Japanese PRTR	http://www.env.go.jp/en/chemi/prtr/prtr.html
Mexico	Registro de emisiones y transferencia de contaminantes	http://app1.semarnat.gob.mx/retc/
Norway	Norwegian PRTR	http://www.norskeutslipp.no
United States	Toxics Release Inventory	http://www.epa.gov/tri

Source: developed by the authors from public websites

PRTRs reach around the world: U.S.'s Toxics Release Inventory (TRI), Canada's National Pollutant Release Inventory, South Korea's Pollutant Release and Transfer Register, Australia's National Pollutant Inventory, Europe's PRTR, and Japan's Pollutant Release and Transfer Register. The rest of this chapter focuses attention on the U.S. TRI and the European PRTR.

United States: Toxics Release Inventory

The U.S. TRI is perhaps the best-known example of this methodology, which is based on the use of information as a regulatory instrument. The Environmental Protection Agency (EPA) developed the TRI, which was promoted by the Emergency Planning and Community Right to Know Act of 1986 (also known as Title III of the Superfund Amendments and Reauthorization Act of 1986). Since 1988, the EPA has focused on installations that manufacture or process more than 25,000 pounds or otherwise use more than 10,000 pounds of any listed chemical or more than 600 toxic chemicals during a calendar year (U.S. EPA 1999). It is credited with reducing chemicals emissions by 45.5 percent in the 1990s and had major repercussions in the media (Khana, Quimio, and Bojilova 1998). TRI also had a negative effect in the share price of firms with poor environmental performance (Hamilton 1995). As a result, firms reacted by significantly reducing their emissions (Konar and Cohen 1997). After the publication of the TRI in the United States, some environmental pressure groups attempted to make the information easily comparable. Rankings on the basis of levels of pollution were developed and published in the mass media (Cañón de Francia, Garcés-Ayerbe, and Ramírez-Alesón 2007).

European Pollutant Release and Transfer Register

In Europe, public access to an inventory of toxic emissions was originally put forward in the European Council Directive 96/61/EC concerning Integrated Pollution Prevention and Control (IPPC). This directive was followed by the European PRTR, which publishes details of individual emissions of 50 classes of toxic substances every three years. Specifically, the European PRTR shows installations' emissions individually and includes all industrial and livestock-sector installations that have acknowledged exceeding the reporting thresholds for one or more of the pollutants listed in European Union Decision 2000/479/CE. These thresholds are not emission limit values, so the data published do not necessarily imply noncompliance with environmental legislation. Since European PRTRs show installation emissions individually, the data are not aggregated. Therefore, it is more difficult to compare them. The EU makes adjustments over time so that installations provide more accurate information about their emissions and environmental practices. Researchers have focused on the importance of the European PRTRs in the environmental sector (Gerde and Logsdon 2001), their influence on share prices in the Spanish stock market (Cañón-de-Francia, Garcés-Ayerbe, and Ramírez-Alesón 2007), and the difficulty in comparing environmental data across different sectors and countries (Sullivan and Gouldson 2007).

PRTRs: Signaling

PRTRs send signals to stakeholders and markets about an installation's environmental performance. Signaling theory suggests that the key attributes of the firm provide information that shapes the impressions individuals have of the organization (Rynes 1991). These key attributes can also be used to examine a firm's reputation and its impact on individual behaviors, attitudes, and decision making (Dutton and Dukerich 1991; Dutton, Dukerich, and Harquail 1994). Firms send signals to those who receive/see these signals (spectators), and the spectators use the signals to form impressions of the firms (Basdeo et al. 2006).

Information, or signals, sent by the firm or other stakeholders can serve as an information-processing shortcut when individuals evaluate decisions concerning the firm. For instance, companies that actively comply with environmental regulations signal that they have some degree of interest in the natural environment (Jones and Murrell 2001), and firms that choose socially responsible actions may signal positive images to higher-quality employees (Fombrun and Shanley 1990). Studies have also shown that consumers are often sensitive to the social performance of companies when making purchasing decisions (Porter and van der Linde 1995). Researchers (Hall 1992; Rindova and Fombrun 1999) state that favorable stakeholder impressions are valuable to firms because they increase stakeholders' willingness to exchange resources with them.

The more information stakeholders have, the easier it is for them to form impressions about a firm, and the better able they are to understand the firm's strategy (Smith and Grimm 1991). Because extremely large quantities of information are provided by PRTRs, it is also important that it be easily accessed and understood by interested stakeholders. To generate information that is easier to understand, PRTRs usually rank installations by summing up the annual emissions of substances released in a given year. Mass media (Kay 2002), management literature (Cohen, Fenn, and Konar 1997; Konar and Cohen 2001, among others), NGOs (IRRC 2002), and other stakeholders often apply this approach.

PRTRs: The Importance of the Toxicity Generated by the Facility

Stakeholders (consumers, environmentalist groups, financial institutions, insurance companies, and investors) are increasingly using PRTR data to measure an organization's environmental performance (Toffel and Marshall 2004). Specifically, community stakeholders are concerned with an installation's environmental information provided by PRTRs because of the consequences for environmental impacts and human hazards. Community stakeholders include geographic communities at large and community groups organized around a political or social cause or interest. The latter may include environmental groups or organizations that "can mobilize public opinion in favor of or against a corporation's environmental performance" (Henriques and Sadorsky 1999: 89)—and are especially likely to do so when such performance influences their welfare. Pressures from such community stakeholders may lead firms to improve their environmental performance (Berry and Rondinelli 1998; Rugman and Verbeke 1998).

It should be noted that PRTRs do not report a firm's environmental emissions per se—they report the activities of specific installations. The installations themselves must report chemical emissions data. Firms send information on each of their emissions to the PRTRs independently of each other, which means that the quantitative data of every hazardous substance generated by each installation are delivered separately. Such data must be carefully interpreted. The toxicity of each pollutant is also very different. Whereas 100,000 kg/year of CO2 is still a legal quantity, more than 1 kg/year of brominated diphenylether can generate negative effects and is illegal. The fact that these pieces of information, which represent the environmental behavior of every firm, are accessible to any agent who may be interested suggests the ways they can be used to influence corporate operations and other actions.

PRTRs: The Importance of Operational Dynamics

The literature states that the firm's size can affect its level of emissions and its desirable environmental behavior (Kerret and Gray 2007). In fact, the concept of eco-efficiency encompasses the importance of the size of the organization. This concept was introduced by the World Business Council for Sustainable Development and defined as follows: "Eco-efficiency is reached by the delivery of competitively priced goods and services that satisfy human needs and bring quality of life, while progressively reducing ecological impacts and resource intensity throughout the life cycle, to a level at least in line with the earth's estimated carrying capacity" (WBCSD 2000: 9). In order to achieve this objective, organizations employ different procedures like environmental management systems and pollution prevention. To implement these procedures, smaller firms must tackle greater challenges than large ones in obtaining and using financial and other scarce resources and capabilities (Hillary 2004).

Consequently, to improve the quality of the data, PRTRs should take into account an installation's size. To account for the installation's size, some adjustments to the data from PRTRs are necessary (Karam, Craig, and Currey 1991), such as including operational dynamics. Operational dynamics play a crucial role in measuring environmental performance (Cairncross 1992; Hart 1995; Schmidheiny 1992). Operational dynamics are strongly related to the scale of a facility's (or firm's) operations: the larger it is, the longer the period of operation, the more it produces—all other things being equal—the more it pollutes. For example, Karam, Craig, and Currey (1991) suggest including operational dynamics activity by considering the number of employees or units produced so that PRTRs measure the environmental performance more accurately.

Therefore, it is useful to know not only the total amount of pollution from installations and firms but also how "efficient" are the installations and firms. The relationship between size and what is produced is important in evaluating how much pollution is being generated. To contribute to such assessments, some PRTRs include data about installations' number of operating hours and number of employees.

Conclusions

Diverse stakeholders try to influence firms to improve their environmental performance. The regulatory stakeholder appears to play a crucial role in determining corporate environmental behavior. In the past decade, governments forced firms to disclose environmental information through PRTRs to reduce the information asymmetry and send signals about corporate environmental performance to society and markets. As a result, governments expect the stakeholders to influence the firms' environmental performance. In the current model, PRTRs provide an installation's environmental information only in the form of raw data and rank installations by summing up annual emissions. The literature states that these governmental efforts are not effective and sufficient if PRTRs are reported only in this format. To improve the effectiveness of PRTRs, Toffel and Marshall (2004) argue that the use of weighting emissions data may provide ranks more accurately. Other researchers propose the need for supplementary information from installations to rank them from an environmentally efficient point of view (Karam, Craig, and Currey 1991). The ways information included in the PRTRs are relevant to and useful for regulators, stakeholders, and managers are not yet well understood and need to be studied far more extensively. However, it is possible to note some of the ways the information is currently being used and to suggest how it might be used more effectively.

First, regulators are paying attention to the total amount of emissions because it is important to understand the specific situations in a given country. This analysis, however, may also be useful to distinguish those firms developing an effort to reduce their environmental impact and to be more efficient. Governments might want to include different size indicators (especially operating hours or production levels) to analyze the relative amount of emissions from each facility.

Second, the information sources of PRTRs may be very important for stakeholders. These registers provide crucial information to the stakeholders to determine on which companies they should put pressure. The PRTR philosophy is based on societal pressure, which works to modify the behavior of those companies that show substandard environmental performance. If information is not accurate, however, stakeholders may focus their efforts on companies that are performing better than other companies that look good on the metrics, but whose environmental performance is actually worse. So it is important to continue to improve the quality of data in the

PRTRs, as well as to provide relatively easy and valid ways to analyze the data and to make comparisons across installations and firms.

From a managerial perspective, the extent to which firms respond to stakeholder pressures becomes a critical concern (Kassinis and Vafeas 2002). Business leaders admit that environmental protection measures have a growing influence on how companies operate (Schmidheiny 1992). Where the PRTR data may be misleading or easily misinterpreted, environmentally responsible firms may have an incentive to present their own recalculations of the information provided in the PRTRs because, in that case, they can signal their good environmental performance more effectively. Environmentally responsible firms may also want to work collectively to contribute to owngoing efforts to improve the quality of the PRTRs' metrics and the data collected and reported in the PRTRs. By taking a positive and proactive stance, environmentally responsible firms may have a major contribution to make to this aspect of global sustainability because of the better access to resources they have. If they take such a stance, and avoid the temptation to attempt to manipulate the standards to minimize and weight lightly the types of pollutants and emissions their operations are most prone to generate, they may stand up well to public scrutiny and build support among important stakeholders (Bansal and Clelland 2004; Meyer and Rowan 1977; Suchman 1995). Such business-based leadership is sorely needed at present and in the continuing future.

References

Agle, R. B., R. K. Mitchell, and J. A. Sonnenfeld. 1999. Who matters to CEOs? An investigation of stakeholder attributes and salience, corporate performance, and CEO values. *Academy of Management Journal* 42(5): 507–525.

Akerlof, G. A. 1970. The market for "lemons": Quality uncertainty and the market mechanism. *The Quarterly Journal of Economics* 84(3): 488–500.

Antweiler, W., and K. Harrison. 2007. Canada's voluntary ARET program: Limited success despite industry cosponsorship. *Journal of Policy Analysis and Management* 26(4): 755–773.

Bansal, P., and I. Clelland. 2004. Talking trash: Legitimacy, impression management, and unsystematic risk in the context of the natural environment. *Academy of Management Journal* 47(1): 93–103.

Baron, D. P. 1995. Integrated strategy: Market and nonmarket components. *California Management Review* 37(2): 47–65.

Basdeo, D. K., K. G. Smith, C. M. Grimm, and V. P. Rindova. 2006. The impact of market actions on firm reputation. *Strategic Management Journal* 27(12): 1205–1219.

Berman, S. L., A. C. Wicks, S. Kotha, and T. M. Jones. 1999. Does stakeholder orientation matter? The relationship between stakeholder management models and firm financial performance. *Academy of Management Journal* 42(5): 488–506.

Berry, M. A., and D. A. Rondinelli. 1998. Proactive corporate environmental management: A new industrial revolution. *Academy of Management Executive* 12(2): 38–50.

Buysse, K., and A. Verbeke. 2003. Proactive environmental strategies: A stakeholder management perspective. *Strategic Management Journal* 24(5): 453–470.

Cairncross, F. 1993. *Costing the earth*. Boston: Harvard Business Scholl Press.

Cañón-de-Francia, J., C. Garcés-Ayerbe, and M. Ramírez-Alesón. 2007. Analysis of the effectiveness of the first European Pollutant Emission Register (EPER). *Ecological Economics* 67(1): 83–92.

Clarkson, M. B. E. 1995. A stakeholder framework for analyzing and evaluating corporate social performance. *Academy of Management Review* 20(1): 92–117.

Cohen, M. 2002. Transparency after 9/11: Balancing the "right-to-know" with the need for security. *Corporate Environmental Strategy* 9(7): 368–374.

Cohen, M. A., S. A. Fenn, and S. Konar. 1997. *Environmental and financial performance: Are they related?* Nashville, TN: Vanderbilt Center for Environmental Management Studies.

Cordano, M., I. H. Frieze, and K. M. Ellis. 2004. Entangled affiliations and attitudes: An analysis of the influences on environmental policy stakeholders' behavioral intentions. *Journal of Business Ethics* 49(1): 27–40.

Dasgupta, S., H. Hettige, and D. Wheeler. 2000. What improves environmental compliance? Evidence from Mexican industry. *Journal of Environmental Economics and Management* 39(1): 39–66.

Delmas, M., and M. W. Toffel. 2004. Stakeholders and environmental management practices: An institutional framework. *Business Strategy and the Environment* 13(4): 209–222.

Dutton, J. E., and J. M. Dukerich. 1991. Keeping an eye on the mirror: Image and identity in organizational adaption. *Academy of Management Journal* 34(3): 517–554.

Dutton, J. E., J. M. Dukerich, and C. V. Harquail. 1994. Organizational images and member identification. *Administrative Science Quarterly* 39: 239–263.

Fombrun, C., and M. Shanley. 1990. What's in a name? Reputation building and corporate strategy. *Academy of Management Journal* 33(2): 233–258.

Foulon, J., P. Lanoie, and B. Laplante. 2002. Incentives for pollution control: Regulation or information? *Journal of Environmental Economics and Management* 44(1): 169–187.

Freeman, R. E. 1984. *Strategic management: A stakeholder approach*. Boston: Pitman/Ballinger.

Gerde, V. W., and J. M. Logsdon. 2001. Measuring environmental performance: Use of the toxics release inventory (TRI) and other U.S. environmental databases. *Business Strategy and the Environment* 10(5): 269–285.

Hall, R. 1992. The strategic analysis of intangible resources. *Strategic Management Journal* 13(2): 135–144.

Hamilton, J. T. 1995. Testing for environmental racism: Prejudice, profits, political power? *Journal of Policy Analysis and Management* 14 (1): 107–132.

Hart, S. 1995. A natural resource-based view of the firm. *Academy of Management Review* 20(4): 986–1014.

Henriques, I., and P. Sadorsky. 1996. The determinants of and environmentally responsive firm: An empirical approach. *Journal of Environmental Economics and Management* 30(3): 381–395.

———. 1999. The relationship between environmental commitment and managerial perceptions of stakeholder importance. *Academy of Management Journal* 42(1): 87–99.

Hillary, R. 2004. Environmental management systems and the smaller enterprise. *Journal of Cleaner Production* 12(6): 561–569.

Hillman, A., and M. Hitt. 1999. Corporate political strategy formulation: A model of approach, participation and strategy decisions. *Academy of Management Review* 24(4): 825–842.

Hoffman, A. J. 2001. *Competitive environmental strategy: A guide to the changing business landscape.* Washington, D.C.: Island Press.

IRRC: Investor Responsibility Research Center. 2002. IRCC's corporate environmental profiles database key to elements. Washington, D.C.: IRRC.

Jones, R., and A. Murrell. 2001. Signaling positive corporate social performance. *Business and Society* 40(1): 59–78.

Karam, J. G., J. W. Craig, and G. W. Currey. 1991. Targeting pollution prevention opportunities using the Toxics Release Inventory. *Pollution Prevention Review,* Spring: 131–144.

Kassinis, G., and N. Vafeas. 2002. Corporate boards and outside stakeholders as determinants of environmental litigation. *Strategic Management Journal* 23(5): 399–415.

Kay, J. 2002. Refineries top polluters on EPA list in Bay area: Discharges taint air, water and land. *San Francisco Chronicle,* 24 May, section A: 11.

Kerret, D., and G. Gray. 2007. What do we learn from emissions reporting? Analytical consideration and comparison of pollutant release and transfer registers in the United States, Canada, England, and Australia. *Risk Analysis* 27(1): 203–223.

Khanna, M., W. Quimio, and D. M. Bojilova. 1998. Toxic release information: A policy tool for environmental protection. *Journal of Environmental Economics and Management* 36(3): 243–266.

Konar, S., and M. A. Cohen. 1997. Information as regulation: The effect of community right to know laws on toxic emissions. *Journal of Environmental Economics and Management* 32(1): 109–124.

———. 2001. Does the market value environmental performance? *The Review of Economics and Statistics* 83(2): 281–289.

Kulkarni, S. 2000. Environmental ethics and information asymmetry among organizational stakeholders. *Journal of Business Ethics* 27(4): 215–228.

Majumdar, S. K., and A. Marcus. 2001. Rules versus discretion: The productivity consequences of flexible regulation. *Academy of Management Journal* 44(1): 170–179.

Meyer, J. W., and B. Rowan. 1977. Institutionalized organizations: Formal structure as myth and ceremony. *American Journal of Sociology* 83(2): 340–363.

Mitchell, R. K., R. B. Agle, and D. J. Wood. 1997. Toward a theory of stakeholder identification and salience: Defining the principle of who and what really counts. *Academy of Management Review* 22(4): 853–886.

OECD (Organization for Economic Co-Operation and Development). 1996. Pollutant Release and Transfer Registers (PRTRs), a tool for environmental policy and sustainable development: Guidance manual for governments. OECD Paris.

———. 2000. Presentation and dissemination of PRTR data: Practices and experiences. Series on Pollutant Release and Transfer Registers No. 3.

———. 2005. Uses of pollutant release and transfer register data and tools for their presentation—a reference manual. OECD Environment Health and Safety Publications Series on Pollutant Release and Transfer Register No. 7.

Porter, M., and C. Van der Linde. 1995. Green and competitive. *Harvard Business Review,* September-October: 120–134.

Rindova, V., and C. J. Fombrun. 1999. Constructing competitive advantage: The role of firm-constituent interactions. *Strategic Management Journal* 20(8): 691–710.

Rugman, A. M., and A. Verbeke. 1998. Corporate strategy and international environmental policy. *Journal of International Business Studies* 29(4): 819–834.

Rynes, S. L. 1991. Recruitment, job choice, and post-hire consequences: A call for new research directions. In *Handbook of industrial and organizational psychology,* ed. M. D. Dunnette and L. M. Hough, vol. 2: 399–444. Palo Alto: Consulting Psychologists Press.

Schmidheiny, S. 1992. *Changing course: A global business perspective on development and the environment.* Cambridge, MA: MIT Press.

Scholz, J., and W. Gray. 1997. Can government facilitate cooperation? An informational model of OSHA enforcement. *American Journal of Political Science* 41 (3): 693–717.

Shaffer, B. 1995. Firm-level responses to government regulation: Theoretical and research approaches. *Journal of Management* 21(3): 495–514.

Sharma, S., and I. Henriques. 2005. Stakeholder influences on sustainability practices in the Canadian forest products industry. *Strategic Management Journal* 26(2): 159–180.

Shrivastava, P. 1995. The role of corporations in achieving ecological sustainability. *Academy of Management Review* 20(4): 936–960.

Smith, K. G., and C. M. Grimm. 1991. A communication information model of competitive response timing. *Journal of Management* 17 (1): 5–23.

Suchman, M. C. 1995. Managing legitimacy: Strategic and institutional approaches. *Academy of Management Review* 20(3): 571–610.

Sullivan, R., and A. Gouldson. 2007. Pollutant release and transfer registers: Examining the value of government-led reporting on corporate environmental performance. *Corporate Social Responsibility and Environmental Management* 14(5): 263–273.

Tietenberg, T. 1998. Disclosure strategies for pollution control. *Environmental and Resources Economics* 11(3): 587–602.

Tietenberg, T., and D. Wheeler. 2001. Empowering the community: Information strategies for pollution control. In *Frontiers of environmental economics,* ed. H. Folmer, H. L. Gabel, S. Gerking, and A. Rose: 85–120. Cheltenham, UK: Edward Elgar.

Toffel, M. W., and J. D. Marshall. 2004. Improving environmental performance assessment: A comparative analysis of weighting methods used to evaluate chemical release inventories. *Journal of Industrial Ecology* 8(1–2): 143–172.

U.S. EPA. 1999. Major findings from the CEIS review of EPA's Toxics Release Inventory (TRI) database. Washington, D.C.: CEIS.

WBCSD: World Business Council for Sustainable Development. 2000. Eco-efficiency: Creating more value with less impact. http://www.wbcsd.org/web/publications/eco_efficiency_creating_more_value.pdf. (accessed January 10, 2010).

Weil, D., A. Fung, M. Graham, and E. Fagotto. 2006. The effectiveness of regulatory disclosure policies. *Journal of Policy Analysis and Management* 25(1): 155–181.

Williamson, O. 1979. Transaction cost economics: The governance of contractual relations. *Journal of Law and Economics* 22: 233–261.

CHAPTER 8

The Adjustment of Corporate Governance Structures for Global Sustainability

Natalia Ortiz-de-Mandojana, Javier Aguilera-Caracuel, and J. Alberto Aragón-Correa

Introduction

Past corporate business scandals such as Enron and WorldCom, and now the recent contributions of corporate actions to the current global financial and economic crisis, have revealed the need for corporate governance reforms even for managing with current business models. To move to business models capable of meeting the need for global sustainability in general and the challenges of climate change in particular will require bold and creative reform of corporate governance structures and practices.

The approach any business takes to deal with the need for global sustainability has important implications in terms of corporate profitability and legitimacy (Buysse and Verbeke 2003; Delmas, Russo, and Montes-Sancho 2007). Whatever approach is chosen will be influenced and supervised by the firm's board (Kassinis and Vafeas 2002). The literature to date, however, has paid little attention to how boards of directors may encourage a sustainability-consistent environmental strategy.

We propose that for companies to move to business models contributing to global sustainability, the adjustment of the corporate governance structures to accommodate climate change concerns is an important step. Specifically, firms may adapt their corporate governance structures to a more environmentally proactive posture, assigning responsibility for climate change to a specific committee.

This chapter includes four sections. First, we discuss the influence of boards on global sustainability issues and how companies can adapt their corporate governance structures to deal with these issues. We then analyze the internal characteristics of firms that have delegated environmental duties to a specific committee. We continue with an examination of how delegating environmental duties to specific committees contributes to dealing with global sustainability concerns. Finally, we provide data illustrating the material in the first three sections. The data come from a sample of 707 firms in North America and western Europe collected with the Carbon Disclosure Project (CDP) Questionnaire 2008 and from the Bloomberg database.

The study offers observations about the development of good governance guidelines and policymakers' regulations on controversial corporate governance and sustainability topics. In addition, we encourage practitioners (especially board members) to create effective structures to deal with global sustainability.

The Influence of Boards on Global Sustainable Issues: Delegating Leadership on Environmental Issues to a Specific Committee

It is widely recognized that boards need to participate more actively in an organization's decisions. Boards of directors must guide and supervise key strategic decisions within the company (Kroll, Walters, and Wright 2008). Setting a strategic environmental policy is one of the most important decisions. A company may voluntarily choose to develop preventive measures to control the company's impact on the natural environment (Hart 1995; Sharma and Vredenburg 1998). Appropriate goals may reduce environmental impacts and improve corporate profitability and legitimacy (Aragon-Correa 1998; Buysse and Verbeke 2003; Delmas, Russo, and Montes-Sancho 2007; Rugman and Verbeke 1998; Sharma and Vredenburg 1998).

Recent works explore several relationships between environmental decisions and some board characteristics (e.g., Berrone and Gomez-Mejia 2009; Kassinis and Vafeas 2002). Kassinis and Vafeas (2002) analyze how board size, the presence of board members with executive duties, and inside ownership factions are associated with the number of environmental litigation cases. Berrone and Gomez-Mejia (2009) did not find that environmental performance had a higher impact on CEO total pay in firms with both environmental pay policy and environmental committees. But boards may also adopt better environmental decisions not only by monitoring an executive team's actions but also by using information, experience, and other cognitive resources (Zahra, Filatotchev, and Wright 2009). Despite these previous studies, several research questions about the influence of boards on a firm's environmental performance are not yet answered.

Not all boards have the same opportunity to influence the firm's natural environmental strategy. It is reasonable to expect that companies assigning responsibility for sustainable issues to a specific board committee (such as an environmental committee) may get more effective action on key environmental matters (Berrone and Gomez-Mejia 2009). Delegating responsibility for these issues to a specific committee may be an important mechanism to improve a firm's contributions to a sustainable world. Research has examined the impact that specific board committees have on a firm's activities. For instance, audit committees are associated with a lower likelihood of financial fraud and the improvement of the information content of earnings (Dechow, Sloan, and Sweeney 1996; Wild 1994). In addition, Huang, Lobo, and Zhou (2009) show that delegating board governance duties to a specific committee improves board effectiveness.

In recent years, the formalization of board committees has been an important development in corporate governance (Tricker 2009). The three principal board committees are the audit committee, the remuneration committee, and the nominating committee (Tricker 2009). A board, however, may create whatever committees it wants to create. Faced with growing demands on a director's time, some boards form an executive committee, a financial committee, or a general-purpose committee to delegate responsibility for handling specific aspects of the board's business (Tricker 2009).

The establishment of board committees has been strongly recommended as a suitable mechanism for improving corporate governance by delegating specific tasks from the main board to a smaller group and harnessing the contribution of nonexecutive directors (Spira and Bender 2004). Board committees can analyze a particular issue in depth and make recommendations to the full board (Kabat 2004). These analyses and recommendations may lead to greater efficiency, expediency, and flexibility (Kesner 1988).

Specific environmental committees are increasing their presence in companies. In 2008, the *Wall Street Journal* noted that about 25 percent of Fortune 500 companies now have board committees overseeing environmental policies, compared with fewer than 10 percent five years ago. These committees typically try to make sure that executives effectively handle conservation efforts, explore new environmentally friendly ventures like wind power, comply with environmental regulations, and manage related business risks (Lublin 2008).

Determinants of Environmental Delegation

Despite the fact that committees overseeing companies' relations to the natural environment are increasing, not all companies have adopted such committees. Therefore, an understanding of why boards may choose to create or not to create such committees and the environmental and profitability results of such decisions is of considerable interest. Firm and board dynamics are likely to affect decisions about delegating environmental issues to a specific committee. This chapter explores two sets of firm internal characteristics that may influence this delegation decision: general firm characteristics and corporate governance characteristics.

1. General Firm Characteristics

Firm characteristics that may be associated with boards' decisions to create an environmental committee include: type of industry, country, financial performance, and size.

Type of Industry

Certain industries are termed "environmentally sensitive" or "dirty" (Gallagher and Ackerman 2000; Mani and Wheeler 1997). Though definitions vary, certain sectors, such as the chemical or petrochemical sector, are normally included on most lists of such industries. Firms that belong to highly polluting sectors generate significant environmental impacts, such as noise, odors, and effluents, and consume high levels of energy and water.

Environmental management literature has focused on industries with a high environmental impact, since the implications of environmental actions are especially relevant in such industries. In addition, firms in highly polluting sectors often face strict environmental regulations, greater attention from the mass media, and strong environmental activism (e.g., Bansal 2005; Berrone and Gomez-Mejia 2009).

Board members may perceive more pressures from the stakeholders in these sectors and be pushed to create responsive environmental governance structures.

Country

Several studies have analyzed the environmental regulations of various countries (e.g., Rugman and Verbeke 1998; Christmann 2004). Some studies suggest that firms locate their most polluting activities in countries with lax environmental regulations (Vernon 1992). Determining the influence of a country's environmental regulations is not simple, however, because a country's institutional profile is complex (DiMaggio and Powell 1983).

Three dimensions are often used to define national institutional profiles: regulatory, cognitive, and normative (Kostova and Zaheer 1999; Scott 1995). Hoffman (1999) applies these dimensions to define a country's environmental institutional profile. The environmental regulatory dimension is measured by a country's environmental legislation. The cognitive (or cultural) dimension is defined in terms of symbols (words, signs, and gestures) as well as cultural rules that guide understanding of the nature of reality and the frames through which that meaning is developed (Zucker 1983). Finally, the normative (or social) dimension encompasses the rules of thumb, standard operating procedures, occupational standards, and educational curricula. These three national dimensions have great repercussions on a firm's environmental strategies.

Because countries' institutional profiles may have many influences on a company, we propose that board members' decisions about adopting adequate environmental structures will be influenced by the environmental profile of the different countries where the firms operate.

Financial Performance

A firm's financial performance may contribute to decisions about delegating environmental issues to specific committees. Since corporate social performance represents an area of high managerial discretion, the initiation of voluntary environmental policies may depend on the availability of excess funds when such policies are seen as costly to the firm. Indeed, there is some evidence to suggest that when managers have more discretionary funds at their disposal, they may be more likely to view environmental issues as opportunities rather than as threats (Bansal 2005; Sharma 2000). Proactive environmental actions help managers and firms increase their perceived control of hazards associated with search and adoption of technological innovations, reducing environmental unpredictability and risk (Sharma 2000).

For these reasons, we argue that the level of financial resources will influence a firm's adoption of advanced environmental practices and the firm's delegation of environmental issues to a specific committee.

Size

Size is also a determinant of environmental conduct (Aragon-Correa 1998; Christmann 2004) and may influence the delegation decision. Research based on samples of large firms has shown that larger organizations are more likely to undertake the most advanced proactive environmental strategies (Aragon-Correa 1998; Russo and Fouts 1997; Sharma 2000). Scholars have argued that environmental sustainability requires accumulation of, and complex interaction among, skills and resources such as physical assets, technologies, and people (Russo and Fouts 1997; Sharma 2000), so the limited resources of small and medium enterprises (SMEs) might make it unlikely that they will adopt such practices (Greening and Gray 1994; Russo and Fouts 1997).

Several studies of SMEs have highlighted their poor level of environmental commitment, describing them as mainly interested in controlling emissions of pollutants to comply with environmental regulations (Williamson and Lynch-Wood 2001). Descriptive studies in several countries, however, have shown that SMEs may successfully implement environmental strategies consistent with the advanced environmental practices of big firms (e.g., Aragon-Correa et al. 2006), including innovations that prevent pollution at the source rather than pollution control at the end of the pipe. Of course, larger firms are likely to have larger and better-funded boards, and that factor might increase the likelihood that larger firms will choose to create environmental committees. On balance, our expectation is that larger firms will be more likely to establish environmental committees.

2. Corporate Governance Characteristics

Among corporate governance characteristics, we propose that CEO duality, the percentage of independent directors (both related with board independence), and board activism can affect decisions.

CEO Duality

Although there have been many calls from institutional investors and in codes of good corporate governance practice for the separation of the roles of chairman and CEO, they are still frequently combined (Tricker 2009).

Research based on whether role separation improves corporate performance and/or corporate value has been contradictory and inconclusive (Kakabadse, Kakabadse, and Barratt 2006). Some studies found that companies in which the same person holds both positions perform better than those with chairman and CEO role separation (Boyd 1995), while others find negative consequences in such firms (Daily and Dalton 1994) or even no effect (Dalton et al. 1998).

The argument in favor of combined roles is that a dynamic enterprise needs just one leader, since spreading leadership duties between two people can lead to conflict (Tricker 2009). Combining the roles can provide a company with stronger leadership internally and externally (Cadbury 2002). Against this argument, Tricker (2009) noted that separation produces a check-and-balance mechanism, avoids the potential for abuse of power concentrated in a single person, and enables the chief executive to concentrate on managing the business while the chairman handles the running of the board and relations with shareholders and other stakeholders (e.g., the government, the regulators, and the media). Higgs (2003) suggests some advantages of separation may not occur in practice, warning that a chairman who was formerly the chief executive of the same company may simply take for granted his inside knowledge and fail as an informational bridge to the nonexecutive directors.

Independent Directors

The study of directors' independence has been one of the most frequently investigated issues in corporate governance literature. In most of the studies, there is a strong perception that independent directors contribute to better governance. The rationale behind this position is that independent boards are more likely to behave according to the shareholders' interest rather than management's interests (Aguilera 2005). Independent directors are also more likely to recognize that their responsibility encompasses more than shareholders; they are conscious of the needs and expectations of various stakeholders (Wang and Dewhirst 1992).

Since independent boards appear to be an important determinant of board decision quality (Tricker 2009) and may be more conscious of the needs and expectations of various constituencies (Wang and Dewhirst 1992), we propose that boards with more independent members are more likely to adapt their structures to the need for global sustainability.

Board Activism

Boards of directors should be active contributors to the performance of their companies. Director primacy is essential to the functioning of corporations because organizations are complex entities with multiple stakeholders, many of them with competing interests (Bainbridge 2008). Conger, Finegold, and Lawler (1998) suggest that board meetings are usually an important indicator of the diligence of a board.

Stock exchange requirements illuminate this point. The most important U.S. stock exchanges provide directives and standards for the boards of companies listed on those exchanges. For example, the New York Stock Exchange (NYSE) states: "To empower non-management directors to serve as a more effective check on management, the non-manager directors of each listed companies must meet at regularly scheduled executive sessions without management" (NYSE Listed Company Manual 303A.01). Although the rule does not indicate how many times per year outside directors must meet, the need to establish some aspects of board independence, and presumably activism, is clear. The NASDAQ and AMEX standards are substantially similar (Brainbridge 2008).

For these reasons, we argue that firms with active boards may be more likely to delegate environmental issues to specific committees.

How Delegating Environmental Duties to a Specific Committee May Contribute to Dealing with Global Sustainability Concerns

Delegating environmental duties to a specific committee may affect environmental decisions, and those decisions may bring real benefits to the organization and society. The data collected for this study explores how the existence of environmental board committees might influence sustainability-related decisions: a firm's environmental reporting and the development of new green products.

Rising critiques of the negative social and environmental impacts of globalization have led firms to become more active in reporting the activities they undertake to prevent negative impacts (Kolk 2008; Sahay 2009). There are many factors driving the increased reporting, and they can differ among countries. The factors include legislation, government guidelines, initiatives that encourage sustainability reporting, and actions of nongovernmental groups and other strong stakeholder groups, as well as individual companies' motivations for reporting (Sahay 2009). Companies producing comprehensive reports show a commitment to preventing negative externalities and improving relationships with stakeholders. In doing so, they seek to improve their reputations.

The response to the challenge of green management has also meant more than pollution prevention for many companies. For some it involves developing new products and services (Marcus and Fremeth 2009). Some companies have not only been able to lower costs and achieve cost leadership by pursuing environmental efficiency, but have also pursued a more focused strategy on the basis of developing "green products" for niche markets (Shrivastava 1995). Introducing new environmental products allows companies to enhance their earnings. They can make big jumps in product development, as innovations occur in such areas as miniaturization, weight reduction, and design for reuse and reparability (Marcus 2009). The search for ways to extract more economic value from fewer natural resources and raw materials can also uncover ways to improve existing products and services and lead to the development of new ones (Marcus and Fremeth 2009).

A specific committee responsible for environmental issues might provide support for managers on environmental issues, ensure the organization's compliance with all aspects of environmental legislation, and raise the organization's attention to environmental matters. By delegating specific environmental tasks to a specific committee, the environmental issues can be studied in depth, and committee members can make comprehensive recommendations to the full board (Kabat 2004; Spira and Bender 2004). These committees typically try to make sure that executives are effective in handling conservation efforts and new, environmentally friendly ventures like wind power, in complying with environmental regulations, and managing related business risks (Lublin 2008). Thus, firms that delegate environmental issues to specific committees are probably more willing to do environmental reporting and develop new green processes and products.

Evidence in North America and Western Europe

Using data from the Carbon Disclosure Project Questionnaire 2008 and the Bloomberg database, we obtained a total sample of 707 firms from 21 countries based in North America and western Europe. We use a dichotomous variable that indicates whether a specific board committee or other executive body has overall responsibility for climate change. The response is directly obtained from the company's response to the Carbon Disclosure Project 2008 (CDP 6) Questionnaire.

During 2008, CDP sent questionnaires to more than 3,000 of the world's largest corporations requesting information on greenhouse gas emissions, the potential risks and opportunities climate change offers, and strategies for

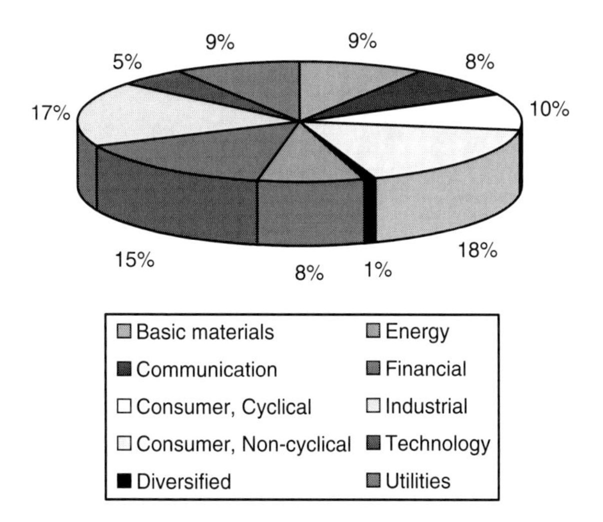

Figure 8.1 Type of industry

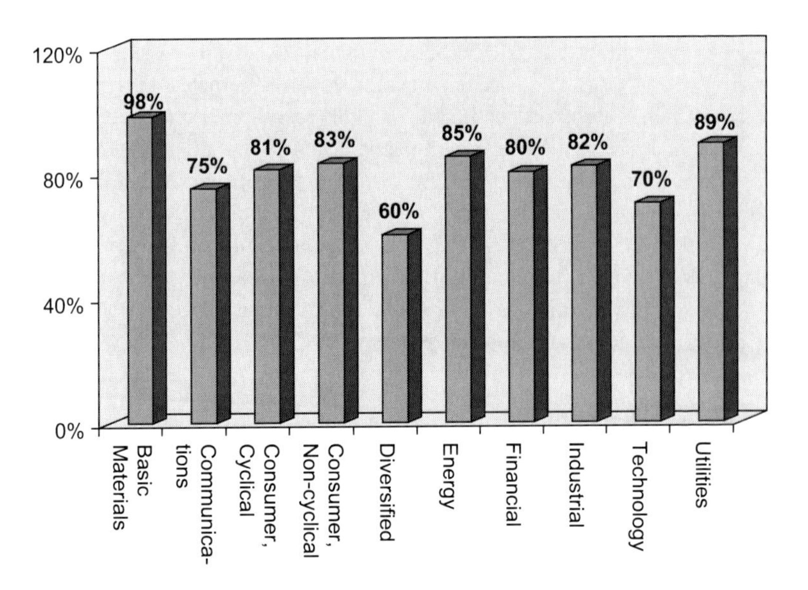

Figure 8.2 Board environmental committees and type of industry

managing those risks and opportunities. CDP received more than 1,550 responses. The overall response rate for Global 500 CDP 2008 was 77 percent. European and North American companies set the pace—with 83 percent and 82 percent response rates respectively. Only 50 percent of Asian Global 500 companies responded (Carbon Disclosure Project 2008).

On the basis of this data, the study reports factors associated with environmental board delegations, and the relationships between the delegation decision and environmental actions.

1. Determinants of the Environmental Delegation

Type of Industry

Using the Bloomberg level I classification of the firms based on their business or economic functions and characteristics, the 707 firms in the sample belong to 9 different industries. Figure 8.1 shows the percentage of firms by sector.

The relationship between the type of industry and a firm's delegation of environmental issues is presented in Figure 8.2. Industries with higher frequencies of companies with environmental board committees include basic materials and utilities. In contrast, firms in diversified industry and technology categories are less likely to have such committees.

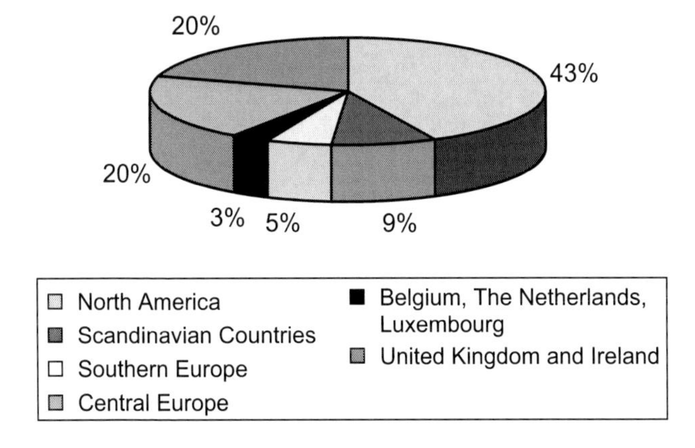

Figure 8.3 Environmental regions

The Chi-Square tests in Figure 8.2 reveal that the differences in terms of having specific committees for environmental issues are statistically significant in relation to the firms belonging to a specific sector.

Firms with specific environmental board committees are the ones that belong to highly polluting industries (basic materials and utilities). In those companies, board members may perceive more pressures from stakeholders and are consequently pushed to create adequate environmental policy structures (Berrone and Gomez-Mejia 2009).

Country

Using World Bank (2001) classifications and the environmental profile of countries, six regions were distinguished: North America (United States and Canada); Scandinavian countries (Sweden, Norway, Finland); Southern Europe (Spain, Portugal, Italy, Greece, Cyprus); The Netherlands, Belgium and Luxembourg; Central Europe (Austria, Germany, France, Denmark, Switzerland); and the United Kingdom and Ireland. Figure 8.3 shows the number of firms headquartered in each environmental region.

The relationship between environmental regions and a firm's delegation of environmental issues to a specific board committee is shown in Figure 8.4. Regions with companies having more environmental board committees include Belgium; The Netherlands and Luxembourg; and the United Kingdom and Ireland. Regions with companies having fewer such committees include Southern Europe and Central Europe.

The Chi-Square test in Figure 8.4 shows that the differences in terms of delegating environmental issues to a specific committee are statistically

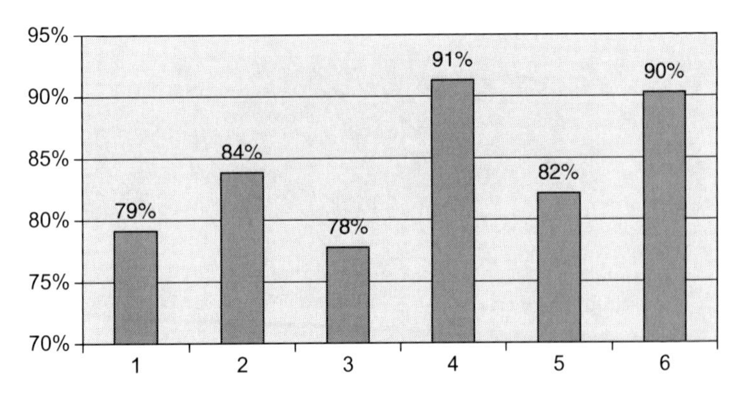

Figure 8.4 Environmental regions and board environmental committees

Table 8.1 Board environmental committees and ROA

	N	*df*	*F Anova*	*Sig*
ROA 2008	705	1	0.05	0.82

Abbreviations: N, the sample size; df, the number of degrees of freedom; F, F-test; Anova, analysis of variance; Sig, statistical significance

significant in relation to the firms belonging to a specific sector. Figure 8.4 shows that firms located in regions with more stringent environmental regulations and advanced environmental institutional profiles are more likely to have environmental board committees.

Financial Performance

Return on Assets (ROA) was used to assess financial performance. This investigation was limited to data for only one year. The Bloomberg database was used to calculate this variable by dividing average assets into trailing 12-month net income (losses) minus trailing 12-month total cash preferred dividends. As Table 8.1 indicates, differences in delegating environmental issues to a specific committee were not associated with the firms' financial performance. In this very limited exploration, the data do not suggest that high levels of financial performance would necessarily imply that firms are more likely to delegate environmental issues to specific committees.

Size

Three indications of firm size were analyzed: "Revenues 2008," which is the total of operating revenues less various adjustments to Gross Sales;

"Employees 2008," which is the total number of company employees at the end of the reporting period as disclosed in the company's social responsibility reports; and "Total Assets 2008," which is the total of all short- and long-term assets as reported on the Balance Sheet.

Differences in terms of delegating environmental issues to a specific committee are statistically significant in relation to the three variables.

Figure 8.5a shows the average number of employees (in thousands) for firms that delegate environmental issues to specific committees and for firms that do not, respectively. Figure 8.5b represents the average of sales (in

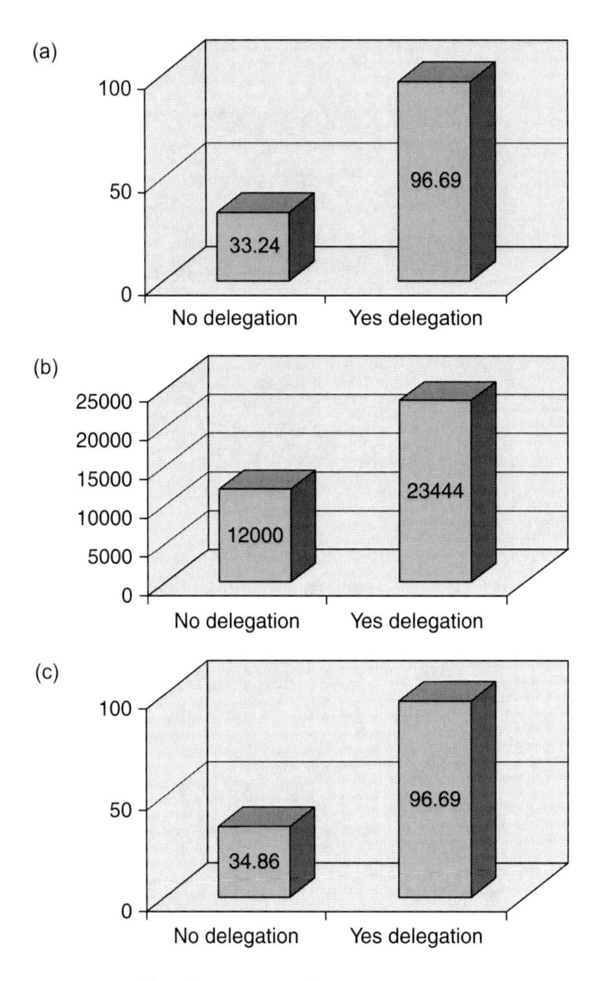

Figure 8.5 Firm size and board environmental committees

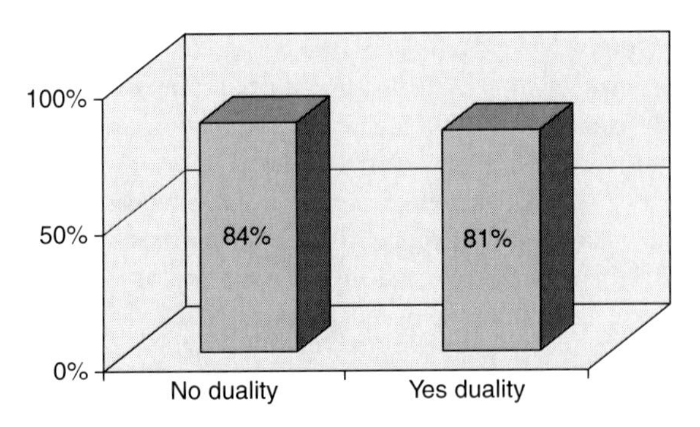

Figure 8.6 CEO-duality and board environmental committees

millions of dollars) for firms that delegate environmental issues to specific committees, and for firms that do not. Finally, Figure 8.5c shows the average assets (in millions of dollars) for firms that delegate environmental issues to specific committees and for firms that do not. These figures show that the largest firms are the ones more likely to establish a separate committee to which they can delegate taking leadership on environmental issues. Firm size greatly favors the delegation of environmental issues to a specific committee.

CEO Duality

The sample consists of 217 firms without CEO duality (59 percent) and 151 firms with CEO duality (41 percent). The relationship between CEO duality and the firms' delegation of environmental issues to a specific committee is represented in Figure 8.6.

According to this figure, firms with the chairman and CEO being different persons have a major percentage of environmental delegation on specific committees. The Chi-Square statistics show, however, that the differences in terms of delegating environmental issues to a specific committee are not statistically significant. Thus, the data do not show an association between CEO duality and delegation of environmental issues to a specific committee.

Independent Directors

Differences in terms of delegating environmental issues to a specific committee are not significant in relation to the firms' percentage of independent directors (see Table 8.2).

Board Activism

To evaluate board activism, the study used the number of board meetings in a firm during 2008 (variable "meeting 2008"). Figure 8.7 shows the

Table 8.2 Independent directors and board environmental committees

	N	*df*	*F Anova*	*Sig*
% Independent directors	437	1	0.526	0.468

Abbreviations: N, the sample size; df, the number of degrees of freedom; F, F-test; Anova, analysis of variance; Sig, statistical significance

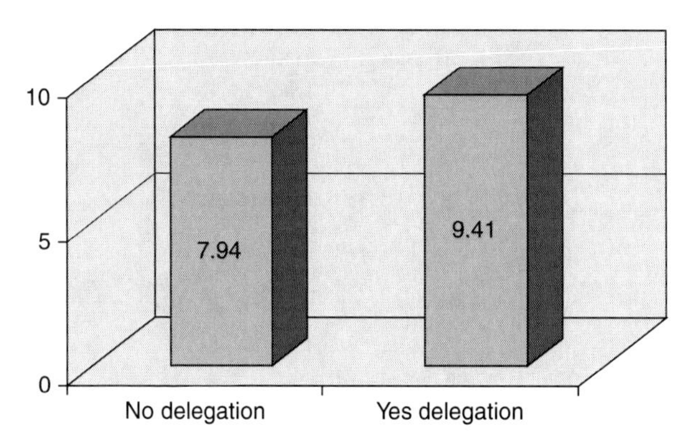

Figure 8.7 Board activism: Board environmental committees and frequency of board meetings

average number of board meetings held in 2008 for firms that delegate environmental issues to specific committees and for firms that do not. Figure 8.7 shows that companies with separate environmental boards met more frequently than companies without such boards.

2. Influence of Environmental Delegation

Environmental Reporting

This variable indicates whether the firm publishes information about the environmental risks and opportunities associated with its operations. The measure is based on companies' reports of their emissions and managerial plans to reduce environmental impacts as reported in their responses to the CDP questionnaire.

According to the available data, 56 firms (22 percent) publish public information about how the firms face climate change concerns, while 202 firms (78 percent) do not publish this type of information.

The relationship between publishing environmental information and a firm's delegation of environmental issues to a specific committee is represented in Figure 8.8a.

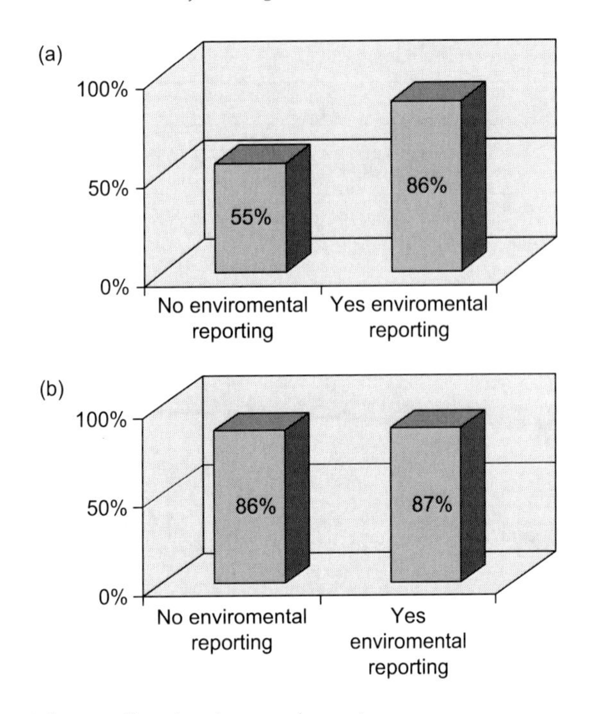

Figure 8.8 Influence of board environmental committees

The Chi-Square test shows that the differences in terms of delegating environmental issues to a specific committee are statistically significant in relation to the firm's announcement of public information about how it faces climate change issues. Thus, firms more likely to do environmental reporting are also ones more likely to delegate environmental issues to a board environmental committee.

Development of New Products

This variable indicates whether the company has developed products in response to climate change. According to the available data, 86 firms (50 percent) reported developing new products on that basis, while 87 firms (50 percent) did not.

The relationship between developing new products and a firm's delegation of environmental issues to a specific committee is represented in Figure 8.8b.

The Chi-Square test shows that the differences in development of new products for firms that do and do not delegate environmental issues to a specific board committee are not statistically significant.

Conclusions

Past corporate scandals such as Enron and WorldCom, and now the recent contributions of corporations to the global financial crisis, have revealed the need for corporate governance reforms for managing even with current business models. The serious and potentially cataclysmic implications of global climate change make the need of new business models consistent with global sustainability even more pressing. To move to models consistent with global sustainability, the adjustment of corporate governance structures to deal with the challenges of climate change will require bold and creative forms of corporate governance reforms.

Delegation of environmental issues to specific committees can contribute to a firm's development of an advanced posture on environmental issues. A specific committee responsible for environmental issues may provide support to managers regarding environmental issues, ensure the organization's compliance with all aspects of environmental legislation, raise the attention paid to environmental matters, and perhaps even contribute to the formulation of corporate strategies consistent with the need for global sustainability.

The type of industry, country, and size of companies are associated with the delegation of environmental issues to specific board committees. Firms in highly polluting industries, regions with stringent environmental institutions, and firms with larger size are more likely to do delegation. In contrast, a firm's financial performance may not have an impact on the delegation of environmental issues to specific committees although the data and analysis used to explore such a relationship were very limited.

In terms of corporate governance characteristics, a firm's board activism was positively related to the delegation of environmental issues to specific committees but no relationship was found between the existence of separate board environmental committees and CEO-duality or percentage of independent directors. Finally, the delegation of environmental issues to specific committees was positively related to environmental reporting but not to development of new "green" products.

Acknowledgments

The research for this chapter was partially funded by the Spanish Ministry of Education and the Andalusian Regional Government (Excellence Research Project P06-SEJ-2356). We want to thank members of the research group ISDE (University of Granada) for their insightful comments.

References

Aguilera, R. V. 2005. Corporate governance and director accountability: An institutional comparative perspective. *British Journal of Management* 16(s1): S39–S53.

Aragon-Correa, J. A. 1998. Strategic proactivity and firm approach to the natural environment. *Academy of Management Journal* 41(5): 556–567.

Aragon-Correa, J. A, N. E. Hurtado-Torres, S. Sharma, and V. J. García-Morales. 2006. Environmental strategy and performance in small firms: A resource-based perspective. *Journal of Environmental Management* 86: 88–103.

Bainbridge, S. M. 2008. *The New Corporate Governance in Theory and Practice*. New York: Oxford University Press.

Bansal, P. 2005. Evolving sustainably: A longitudinal study of corporate sustainable development. *Strategic Management Journal* 26(3): 197–218.

Berrone, P., and L. R. Gomez-Mejia. 2009. Environmental performance and executive compensation: An integrated agency-institutional perspective. *Academy of Management Journal* 52(1): 103–126.

Boyd, B. K. 1995. CEO duality and firm performance: A contingency model. *Strategic Management Journal* 16(4): 301–312.

Buysse, K., and A. Verbeke. 2003. Proactive environmental strategies: A stakeholder management perspective. *Strategic Management Journal* 24(5): 453–470.

Cadbury, A. 2002. *Corporate Governance and Chairmanship: A Personal View.* Oxford: Oxford University Press.

Carbon Disclosure Project. 2008. Global 500 Report. PricewaterhouseCoopers LLP (PwC), www.cdproject.net (accessed April 30, 2010).

Christmann, P. 2004. Multinational companies and the natural environment: Determinants of global environmental policy standardization. *Academy of Management Journal* 47(5): 747–760.

Conger, J. A., D. Finegold, and E. E. Lawler, III. 1998. Appraising boardroom performance. *Harvard Business Review* 76(1): 136–148.

Daily, C. M., and D. R. Dalton. 1994. Bankruptcy and corporate governance: The impact of board com. *Academy of Management Journal* 37(6): 1603–1617.

Dalton, D. R., C. M. Daily, A. E. Ellstrand, and J. L. Johnson. 1998. Meta-analytic reviews of board composition, leadership structure, and financial performance. *Strategic Management Journal* 19(3): 269–290.

Dechow, P. M., R. G. Sloan, and A. P. Sweeney. 1996. Causes and consequences of earnings manipulations: An analysis of firms subject to enforcement actions by the SEC. *Contemporary Accounting Research* 13(1): 1–37.

Delmas, M., M. V. Russo, and M. J. Montes-Sancho. 2007. Deregulation and environmental differentiation in the electric utility industry. *Strategic Management Journal* 28(2): 189–209.

DiMaggio, P., and W. Powell. 1983. The iron cage revisited: Institutional isomorphism and collective rationality in organizational fields. *American Sociological Review* 48(1): 147–160.

Gallagher, K., and F. Ackerman. 2000. Trade liberalization and pollution intensive industry in developing countries: A partial equilibrium approach. Meldford, MA: Tufts University, Global Development and Environment Institute.

Greening, D. W., and B. Gray. 1994. Testing a model of organizational response to social and political issues. *Academy of Management Journal* 37: 467–498.

Hart, S. L. 1995. A natural-resource-based view of the firm. *Academy of Management Review* 20(4): 986–1014.

Higgs, D. 2003. *Review of the Role and Effectiveness of Non-Executive Directors.* London: DTI.

Hoffman, A. J. 1999. Institutional evolution and change: Environmentalism and the U.S. chemical industry. *Academy of Management Journal* 42(4): 351–371.

Huang, H., G. Lobo, and J. Zhou. 2009. Determinants and accounting consequences of forming a governance committee: Evidence from the United States. *Corporate Governance: An International Review* 17(6): 710–727.

Kabat, B. 2004. Strengthening board-manager relations. *Management Quarterly* 45(3): 50–55.

Kakabadse, A., N. K. Kakabadse, and R. Barratt. 2006. Chairman and chief executive officer (CEO): That sacred and secret relationship. *The Journal of Management Development* 25(2): 134–150.

Kassinis, G., and N. Vafeas. 2002. Corporate boards and outside stakeholders as determinants of environmental litigation. *Strategic Management Journal* 23(5): 399–415.

Kesner, I. F. 1988. Directors' characteristics and committee membership: An investigation of type, occupation, tenure, and gender. *Academy of Management Journal* 31(1): 66–84.

Kolk, A. 2008. Sustainability, accountability and corporate governance: Exploring multinationals' reporting practices. *Business Strategy and the Environment* 17(1): 1–15.

Kostova, T., and S. Zaheer. 1999. Organizational legitimacy under conditions of complexity: The case of the multinational enterprise. *Academy of Management Review* 24(1): 64–81.

Kroll, M., B. A. Walters, and P. Wright. 2008. Board vigilance, director experience, and corporate outcomes. *Strategic Management Journal* 29(4): 363–382.

Lublin, J. S. 2008. Theory and practice: Environmentalism sprouts up on corporate boards; more companies start panels on green issues amid push by activists. *Wall Street Journal* (Eastern edition). New York: Aug 11, 2008. p. B.6.

Mani, M., and D. Wheeler. 1997. *In Search of Pollution Havens? Dirty Industry Migration in the World Economy.* Washington, D.C.: World Bank.

Marcus, A. 2009. Strategic directions and management. In *Business, Management and Environmental Stewardship,* ed. Robert Staib. New York: Palgrave Macmillan.

Marcus, A., and A. Fremeth. 2009. Green management matters regardless. *Academy of Management Perspectives* 23(3): 17–26.

Rugman, A. M., and A. Verbeke. 1998. Corporate strategies and environmental regulations: An organizing framework. *Strategic Management Journal* 19(4): 363–375.

Russo, M.V., and P. A. Fouts. 1997. A resource-based perspective on corporate environmental performance and profitability. *Academy of Management Journal* 40: 534–559.

Sahay, A. 2009. Organization: Structures, frameworks, reporting. In *Business, Management and Environmental Stewardship,* ed. Robert Staib. New York: Palgrave Macmillan.

Scott, W. R. 1995. Institutions and organizations. London: Sage Publications.

Sharma, S. 2000. Managerial interpretations and organizational context as predictors of corporate choice of environmental strategy. *Academy of Management Journal* 43: 681–697.

Sharma, S., and H. Vredenburg. 1998. Proactive corporate environmental strategy and the development of competitively valuable organizational capabilities. *Strategic Management Journal* 19(8): 729–753.

Shrivastava, P. 1995. The role of corporations in achieving ecological sustainability. *Academy of Management Review* 20(4): 936–960.

Spira, L. F., and R. Bender. 2004. Compare and contrast: Perspectives on board committees. *Corporate Governance: An International Review* 12(4): 489–499.

Tricker, B. 2009. *Corporate Governance: Principles, Policies and Practices.* New York: Oxford University Press.

Vernon, R. 1992. Transnational corporations: Where are they coming from, where are they headed? *Transnational Corporations* 1(2): 7–35.

Wang, J., and H. D. Dewhirst. 1992. Boards of directors and stakeholder orientation. *Journal of Business Ethics* 11(2): 115–123.

Wild, J. J. 1994. Managerial accountability to shareholders: Audit committees and the explanatory power of earnings for returns. *The British Accounting Review* 26(4): 353–374.

Williamson, D., and G. Lynch-Wood. 2001. A new paradigm for SME environmental practice. *TQM Magazine* 13: 424–433.

World Bank. 2001. World Development Report. New York: Oxford University Press.

Zahra, S. A., I. Filatotchev, and M. Wright. 2009. How do threshold firms sustain corporate entrepreneurship? The role of boards and absorptive capacity. *Journal of Business Venturing* 24(3): 248–260.

Zucker, L. 1983. Organizations as institutions. In *Research in the Sociology of Organizations,* ed. S. Bacharach: 1–47. Greenwich, CT: JAI Press.

CHAPTER 9

Taxes and Sustainability

John L. Stancil

Adam Smith is generally regarded as the godfather of modern economics. His four principles of Equity, Certainty, Convenience, and Efficiency remain the cornerstone of tax policy analysis today. But the tax landscape has broadened exponentially in the years since the publication of *The Wealth of Nations*. Today, most authorities would list far more principles than Smith's four. When Adam Smith postulated his principles of good tax policy, he probably had no conception of the extent to which taxes would be utilized as instruments of economic and social policy. He even stated that the tax system should not attempt "social engineering." According to Smith, a tax system should not attempt to encourage or discourage certain types of behavior (McHugh 2004).

Today, the tax code is widely accepted as an instrument of economic and social policy. Many times, the pursuit of social and economic goals is the overriding factors in tax policy, while the raising of revenues takes on secondary importance. In the arena of environmental taxation, credits or deductions for energy-efficient expenditures are commonplace. Congestion fees are levied to discourage use of highways during periods of heavy use (*Science Daily* 2008). Taxes on natural resources have been levied to help reduce consumption. Carbon and sulfur emissions have been the subject of tax levies (Fischlowitz-Roberts 2002). Frequently, these taxes are not meant to increase governmental revenues as they are implemented along with other actions that decrease income or other taxes. Such changes are often seen as having an added benefit of reducing distortions in the economy (tutor2u 2009).

The area of taxation for sustainability is known by a number of names: green taxation, eco-taxation, energy incentives, or environmental taxation. Each of these concepts, while focused on similar objectives, seems to lack a macroview of the subject. Green is a popular word today, but the term in relation to the environment seems to be a fad. Eco-taxation suffers from a lack of identity. Does "eco" stand for economic, ecology, or some other term? The term "environmental taxation" has been defined as a tax aiming to ensure that polluters face the true cost of their activities by charging them for the damage done to others (Tax Policy Center 2008). This approach is more of a "stick" approach and offers no "carrot" to encourage environmental stewardship. Energy measures do not cover the entire area of taxation, but tend to focus on the incentive side. One term that has not received a great deal of usage, but seems to be superior to the others is "sustainability taxation." This term includes environmental taxation but takes a larger view, a view toward doing what is necessary to assure a sustainable future for this world. This term can encompass taxes, fees, or other measures that encourage businesses and consumers to move from less-desirable environmental actions to those that can help create a more sustainable future. Olivia Sprinkel (2009) defines sustainability as "a balance between the financial, human, and environmental."

In the sustainability arena, the term taxes is not normally restricted to the classic definition of a tax, but encompasses any charge or fine levied by a governing authority that seeks to promote a sustainable lifestyle in society. Taxes, fines, charges, and tariffs all come under the sustainability umbrella.

Origins of Sustainability Taxation

The concept of sustainability taxation probably was originated in 1920 by the economist A. C. Pigou. Pigou drew a distinction between the private and the social value of economic activities. A modern illustration of this principle would be the construction of a new toll road. The users of the highway enjoy the private benefits of the new road: reduced congestion, quicker trips, and the like. The benefits are reflected in the price users pay to travel the new route. But at the same time, there are social costs. People are displaced as the new road cuts through their neighborhoods. There is an increase in noise from the traffic. Pollution increases. These social costs, or externalities, do not enter into the calculations of the cost of the road but must be included in determining the ultimate worth of any economic activity. To correct these problems, Pigou advocated government intervention. Where the social value of an activity was less than its private value, the authorities should introduce "extraordinary restraints" in the form of user taxes. Pigou also realized that some activities have a social value exceeding the private value. Recreational parks, street lamps, and other "public goods" are difficult projects to charge for, so the free market would not ensure an adequate supply. Pigou suggested "extraordinary encouragements" in the form of government subsidies to help assure an adequate supply of these "public goods" (Cassidy 2009). Pigou's theories form the foundation of today's concept of sustainability taxation.

Why Use Taxation in the Environmental Policy Mix?

It has become apparent that a successful approach to achieving a sustainable future will involve a mix of policy initiatives. Recycling, use of renewable energy sources, new technologies, and other measures will move this planet toward a sustainable future. But there must be in place an incentive to impel consumers and businesses to implement such sustainable actions. Brechling has argued that the pricing mechanism, in a free-market economy, is likely to be lacking in at least three respects:

- The overall price elasticity of demand for energy is low and the level of taxation on energy to induce substantial behavioral change will be too high to be acceptable to many.
- The regressive nature of environmental taxes will have negative effects on wealth distribution, as low-income groups are affected in a disproportionate way.
- There may be various obstacles or "market failures," which prevent efficient levels of energy-efficient investments.

On the first point, Noriko Fujiwara, Jorge Nunez Ferrer, and Christian Engehofer (2006) observe that elasticities of demand for environmental taxes are very important. If elasticity of demand is high, rapid and successful implementation of such taxes is possible. Such an example would be the implementation of a tax on plastic bags. Consumers use these because they are convenient and have no visible cost for their use. Hence, there is no financial benefit from not using them, and a switch to reusable bags (a viable substitute) carries a cost to the consumer. However, a tax on the plastic bags creates a cost with an incentive to invest in reusable bags.

On the other hand, the situation is different for carbon taxes. Elasticity of demand for energy is low in the short term and energy is an important input for large sections of the economy. Additionally, there are currently no reliable, low-cost alternatives to fossil fuels. Energy users cannot easily reduce consumption or switch to alternatives when the price of energy rises (European Commission 2000). This short-term elasticity and the lack of alternatives create problems for carbon taxes. However, Fujiwara, Ferrer, and

Engehofer observe that this situation does not eliminate the need for carbon taxes, but means that taxation needs particular care to be effective without causing adverse effects. The methods, implementation, and structure need special attention.

Sustainability taxes do have the potential to be regressive, falling in a disproportionate measure on those in the lower income brackets. Behavioral changes are the ultimate goal of such taxes, but these changes take place over the long term. Short-term implications should not be ignored, as these will precede any behavioral changes. In enacting any sustainability taxes, analysis of the impact of the taxes must pinpoint which sectors of society are hit the hardest and which will not be able to adapt to the change (Fujiwara, Ferrer, and Engehofer 2006).

One tendency is to attempt to mitigate the regressive aspects by initiating compensation measures for those in the most severely impacted categories through tax exemptions or refundable credits to compensate for the additional burden imposed by the tax. The Organisation for Economic Co-operation and Development (OECD) (2001) recommends against mitigation measures such as reduced tax rates for low-income households in addressing regressivity concerns, as they may reduce the incentive for behavioral change. They suggest other unspecified means of reducing the impact. These alternative measures would be intended to reduce the undesirable impact of the sustainability tax while maintaining the price signal of the tax. Maintenance of the price signal keeps intact the incentives to modify behavior in an environmentally beneficial manner.

The efficiency of these taxes, the third concern raised by Brechling, remains an uncertain area. Energy taxes, by and large, are input taxes and should fall on production as well as consumption. To avoid distortion in production, the tax, however, should be limited to final consumption. This approach is a less-expensive tone for collecting the tax (Newbery 2001). However, it does not provide any incentive for the producer to avoid negative environmental externalities. Inclusion of exemptions, revenue recycling, or other approaches in an attempt to minimize the regressive effect can raise the administrative costs and render the tax an economically inefficient one.

Double Dividends as Sustainable Tax Policy

Proponents of sustainable taxes have often argued that they are "fiscally neutral," meaning that new environmental taxes would be offset by decreases in existing taxes, often ones related to payroll. Citing the best of both worlds, a "double-dividend" has been declared to exist for environmental taxes (Fujiwara, Ferrer, and Engehofer 2006). The first dividend relates to environmental improvements and the second dividend comes as payroll taxes are reduced. Lower payroll taxes have been seen as an improvement in macroeconomic policy. Another view of the double dividend states that the sustainable taxes would reduce pollution and generate health, ecological, and other benefits while eliminating losses associated with other distorting taxes (Morganstern 1995). Unfortunately, the "double dividend" effect has not been empirically demonstrated and there is evidence that the idea may not hold up to detailed analysis. While a modest tax shift has been demonstrated, that shift is not seen as validating the double dividend theory (Morgenstern 1995). One possible reason for this failure to show the double dividend effect may be that the fiscal neutrality approach has given lobbyists an opportunity to seek generous exemptions for well-connected special interest groups under the banner of seeking fiscal neutrality. These lobbying efforts frequently create adverse effects on environmental effectiveness (Fujiwara, Ferrer, and Engehofer 2006).

A related topic is "revenue recycling." Under this concept, funds obtained through taxes or levies on environmental pollution are "recycled" as credits for specific purposes that generate environmental benefits. For example, a credit for the installation of energy-saving investments could be paid from funds obtained from taxes on environmental pollution. This approach depends on effective government actions. The government must allocate and recycle these revenues efficiently and avoid creating distortions and transaction costs (Fujiwara, Ferrer, and Engehofer 2006). Using earmarks in the legislative process and inefficiencies in the governmental bureaucracy can be sources of efficiency losses.

In the United Kingdom, revenues from sustainability taxes have been used to reduce the rate of employers' National Insurance Contributions. Grants have also been made to support research and development projects, interest-free loans, and funding for carbon emissions reductions (Fujiwara, Ferrer, and Engehofer 2006). Such use of funds raises the question of whether these actions are the most efficient uses of the funds because they involve the government bureaucracy in decisions regarding the allocation of the revenues.

Today's Use of Sustainable Tax Measures

Nations have a great deal of diversity in their approach to implementing sustainable tax measures. As discussed earlier, Pigovian taxes are seen by many as the ideal approach to per unit taxes on emissions or discharges. However, these taxes have seen limited use. Outside of Europe, no nation has adopted the Pigovian approach. However, the 30 signatories to the

OECD have utilized indirect environmental levies that include taxes on fuels, vehicles, beverage containers, fertilizers, and other environmentally harmful products or activities. These levies are growing in importance in OECD nations (Barde 2005).

CO_2 taxes are growing in importance—West-European nations have implemented some form of this tax. The effectiveness of these taxes has been limited due to differing systems in each country. Ivan Hodac, Secretary General of the European Automobile Manufacturers' Association (ACEA) stated that CO_2 taxes are important in shaping consumer demand toward fuel-efficient vehicles. However, he called for a harmonized scheme "to give a clear market signal which will be decisive in achieving the desired cuts in CO_2 emissions." He further stated that fragmented systems have a distorting effect on the internal market (AECA 2008).

The Spanish Case

The Spanish corporate income tax includes a tax credit for environmental investments. This credit was originated in 1996 and had a narrow focus. It was limited to investments consisting of installations used to avoid air pollution from industrial facilities; prevent the pollution of surface underground water and seawater; and reduce, recover, or treat industrial waste. Initially, this tax was to be in force for only one year; however, several amendments to the law were enacted, expanding the coverage and the time frame. The tax has now been extended indefinitely and has been expanded to purchases of new land-based means of transportation for commercial or industrial use. Investments related to the use of renewable energy sources are also included in the legislation.

The most recent amendment to this legislation provided a clear incentive to adopt cutting-edge sustainability measures that "go beyond what is legally required" (Ventosa 2008). The credit is 10 percent of the total investment in installations used to (1) avoid air pollution from industrial facilities; (2) prevent pollution of surface, underground, and seawater; or (3) reduce, recover, or adequately treat industrial waste. As mentioned, the investment must go beyond what is legally required and be included in agreements with the relevant environmental authorities. These authorities must issue a validating certificate, but unfortunately this provision has not yet been applied in practice.

The amendment includes 12 percent credit for purchases of new land-based means of transportation for commercial or industrial use, but only the portion of the investment that effectively contributes to the reduction of pollution is eligible for the credit. A second 10 percent credit is available for investments in new tangible assets for the use of renewable energy sources. These investments include:

- use of solar energy for transformation into heat or energy;
- use of municipal solid waste, biomass from agricultural and forestry waste, or biomass from energy plantations to produce fuel;
- treatment of biodegradable waste from livestock activities, water treatment plants, industrial effluents, or municipal solid waste for biogas transformation; and
- treatment of agricultural or forestry product, or used oils for their transformation into biofuels (Ventosa 2008).

Low Carbon Recovery in the United Kingdom

The United Kingdom has become the world leader in approaching the problem of climate change, adopting a strategic, long-range focus. Prior to 2009, the UK had made significant strides toward reducing carbon emissions. Existing policies are supporting £50 billion in low-carbon investments through 2011. Additionally, these policies have supported 900,000 jobs. Budget 2009 provided over £1.4 billion of additional targeted support in the low-carbon sector. Other measures promise an additional £10.4 billion of low-carbon sector and energy investments over the next three years. These measures place the UK at the forefront of worldwide low-carbon recovery. Budget 2009 sets forth the world's first carbon budgets as required in the recently enacted "Climate Change Act." There is a legally binding reduction of 34 percent reduction in emissions by 2020 (HM Treasury 2009).

The £50 billion of energy investments includes £8.9 billion for energy-efficiency measures to help households, businesses, and the public sector use less energy. It includes a reduction in the value-added tax (VAT) for energy-saving materials as well as incentives for thermal insulation in industrial installations (HM Treasury 2009). Six billion pounds are provided for supporting the development and use of renewable energy sources, which could yield a tenfold increase in renewable energy. These funds will not only support increased use of energy from small-scale energy resources such as solar power and heat pumps, but also community heating systems. Those systems generate heat at a central location and transmit the heat via pipes. The development and deployment of low-carbon technologies is the focus of a £1.7 billion appropriation.

These and other Budget 2009 initiatives are not tied to a specific tax resource, but are a reflection of the importance the British place upon achieving a low-carbon future. In other energy-efficient schemes, £365 billion are planned with the intent of reducing emissions, saving money, and helping employment (HM Treasury 2009).

It is apparent that the British sustainability initiatives go beyond the use of taxes and incentives, and looks toward achieving additional revenues through an economic boost resulting from implementation of sustainability measures.

The United States and Sustainable Taxes

The United States has lagged behind its European counterparts in attempting to create a sustainable environment, particularly in regard to utilizing the tax structure to help implement effective sustainability policies. Four federal laws enacted since early 2008 contain provisions targeting energy conservation: the Economic Stimulus Act of 2008, the Housing Assistance Act of 2008, the Emergency Economic Stimulus Act of 2008, and the American Recovery and Reinvestment Act of 2009 (Watson 2009). None of these statutes can be classified as sustainability or "green" legislation, although they do contain certain elements of energy-efficient legislation. The environmental focus in each of these acts is on credits for energy-efficient buildings or building improvements. While this is a laudable move toward sustainability, it can hardly be expected to create a sustainable future for the United States.

This recent flurry of tax legislation is merely a continuation of Federal environmental tax policies that have focused on tax credits and deductions having positive environmental effects rather than sending negative price signals for environmentally damaging activities. The Energy Policy Act of 2005, like other federal legislation, relied heavily on tax incentives for energy conservation investments. Included were incentives for energy-efficient heating, cooling, and lighting systems in commercial buildings; income tax credits for alternative-fuel vehicles; incentives for alternatives to coal-burning plants; and credits for wind farms producing electricity produced from wind power (Milne 2007).

Congestion Pricing

Within the United States, several of the states have taken limited steps toward using taxes to implement sustainability policies. Some states such as Georgia, California, and Virginia are transforming some high-occupancy vehicle (HOV) lanes into high-occupancy toll (HOT) lanes. These systems typically allow vehicles with three or more occupants to travel the designated lanes free of charge, but allow others to use the lanes for a fee. Despite charges that these are "Lexus Lanes" for the wealthy, research in California has shown that the demographics for users of the HOT lanes are similar to those driving the regular lanes (Dodd 2009). HOV lanes have apparently not reduced the number of vehicles on the road, but are used by those who would be traveling together in any case. HOT lanes do not seem to have any significant mechanism for energy conservation, but are merely revenue-raising devices that may reduce traffic congestion slightly (Dodd 2009).

Similarly, the concept of congestion taxes has been utilized in a number of nations in a variety of circumstances. The concept has been applied to waterways, airports, and city-center hubs in addition to highways. Transportation Alternatives called congestion pricing the most powerful policy tool available to New York City officials to reduce unnecessary driving, promote environmentally sound transportation, and finance twenty-first century improvements to the transportation infrastructure. If the revenues are utilized for this purpose, environmental benefits could become a reality. However, Owen (2009) takes a different view. While agreeing that congestion taxes may reduce traffic and provide funds for environmentally friendly public transportation alternatives, he states that there is dubious environmental value to this approach, stating that congestion maintains a level of frustration that turns drivers into pedestrians or subway riders. In this view, congestion pricing has its advantages, but one of them does not seem to be a contribution toward a sustainable future any more than an income tax used for environmental purposes can be said to be a sustainability tax.

Plastic Bag Taxes

The plastic bag has become ubiquitous in our society as we use an estimated 500 billion of these bags annually (ReusableBags.com). However, they are coming under increasing controversy. They are not biodegradable; they kill an estimated 100,000 marine animals annually; and they consume fossil fuels in their manufacture. Additionally, there are viable alternatives.

With these considerations in mind, a number of nations have implemented a tax on each bag. Several cities and states in the United States are considering such proposals. The tax can run from 5 to 33 cents a bag. The high-end of that tax range creates a strong disincentive for their use. If the consumer opts for an alternative, such as a reusable cloth bag, the retailer will purchase fewer bags with the end result that fewer plastic bags will be produced. Paper bags are also not environmentally friendly. Although they do degrade, Rosenthal (2008) reports that they cause the release of more greenhouse gases in their manufacture and transportation than plastic bags do.

Although some cities and towns in the United States have banned plastic bags entirely (Carlson 2008, Roach 2008), the approach of taxing, rather than banning, plastic and paper bags seems to be the promising path to follow. While proponents may desire that every nation ban these bags, the lack of a universal ban on such bags does not diminish the local effect. There may be certain instances where plastic is preferable with no equivalent substitute. Placing products such as meat or fresh fruits and vegetables in plastic or paper bags is one such possible instance. Protecting a product from the elements might be another. While the "bag tax" seems to be effective in reducing the consumption of plastic and paper bags, governments must be judicious in how they approach the use of the resulting tax revenues. Since the goal is to eliminate the use of such bags, the revenue stream can be expected to decrease rather quickly over time. When Ireland introduced its 33 cent tax per bag, consumption decreased 94 percent in a matter of weeks (Lee 2009). This tax, then, is not one designed to bring in revenues, but to change behavior.

A Future View of Sustainability and Taxes

We live in a global society where national borders are easily and frequently transcended. Any tax that seeks to promote sustainability in one nation will only be as effective as taxes enacted in other nations. Companies faced with some aspect of environmental tax regulation will do cost-benefit analyses. Is it more advantageous for the company to remain in its present location and pay the tax, or can the company benefit from moving its operations to another nation where there is a lower level of environmental regulation? Obviously, one partial solution to promote sustainability would be for the "taxing" nation to include tariffs on imports of products manufactured in nations lacking the level of environmental regulation of the "taxing" country if the web of international trading agreements allows it to do so. This approach, however, is likely to result in a sustainable tax policy that is a patchwork of assorted laws and regulations, needing adjustment when other nations amend their sustainable tax policies. It would likely result in an ineffective global sustainable tax policy with resultant gaps and distortions.

Building on the Kyoto Protocol

What is needed for an effective sustainable tax policy is a global approach not unlike the existing Kyoto Protocol. While the Kyoto Protocol is an international treaty designed to bring nations together to reduce climate change and the resulting global warming, a similar treaty could be implemented to coordinate a global approach to sustainable tax policy. Obviously, not all nations would sign and ratify the treaty, the United States being the most notable holdout on ratification of such agreements. The Kyoto Protocol, for example, was signed by President Clinton, but never submitted to the Senate for ratification (Marcoux 2006). Without the participation of the world's two largest producers of greenhouse gases (China and the United States), the future of the Protocol is in doubt past its 2012 expiration date. However, this situation can be an opportunity to craft a new, comprehensive approach to sustainability, incorporating the taxation tool.

Whether a part of a new Kyoto Protocol or some other approach, such a treaty must have certain characteristics to promote sustainability effectively worldwide. Obviously, it cannot move toward global sustainability without the cooperation of all major industrialized nations. Such cooperation is of utmost importance. No matter what provisions are included in the treaty, or how effectively they are seen to promote sustainability, the effectiveness of the treaty will suffer without full-scale participation by all major nations.

There are seven characteristics that must be addressed in a global tax sustainability effort: neutrality, comprehensiveness, coordination, a Pigovian approach, removal of subsidies, social equity, and visibility.

Neutrality

Neutrality will be addressed first, as it is a characteristic that should *not* be sought in attempts to achieve a sustainable tax policy. Neutrality is a widely accepted concept in principle, but one that is frequently not attained due to tradeoffs between different concepts of neutrality and differing goals (Furman 2010). Tax neutrality is generally defined as a tax that does not cause entities to shift economic choice among alternatives. However, the goal of sustainability tax policy is to get entities to change their economic and social behaviors. Therefore, policymakers frequently depart from seeking to meet the neutrality criterion to achieve specific goals. These goals can be two-sided: encouraging desirable activity and discouraging undesirable activity. Because the objective is to alter behavior to achieve sustainability, a sustainable tax policy should not be neutral under this concept of neutrality.

Revenue neutrality is a second framing of the neutrality concept. Many who advocate a sustainable tax policy seek a revenue-neutral policy. The widely publicized double dividend is interpreted by some to contribute to revenue neutrality. Although some tax shifting may appear to occur, the double dividend has not been shown to hold up to a close analysis. Morgenstern states that while environmental taxes do not provide a free lunch, they are a relatively economical approach to addressing sustainability. Environmental benefits associated with a tax shift are generally not costless.

Another consideration in a revenue neutrality analysis is that when tariffs, fees, or other taxes are levied to discourage certain behaviors, those behaviors will decline if those measures are effective, or perhaps for some other reasons. When the revenue from those decreasing behaviors declines, the neutrality of the sustainable tax policy will be upset.

Comprehensiveness

While the need for a comprehensive sustainable tax policy has been addressed in relation to the need to have all major industrialized nations as participants, there is a second aspect to comprehensiveness. This step is probably the most difficult of the characteristics to achieve. A comprehensive sustainable tax policy approach must address all major aspects of sustainability. Failure to do so will result in gaps that nations, companies, and individuals may exploit. There are at least five considerations in forming a comprehensive sustainable tax policy.

First, the policy should contain a commitment raising awareness of sustainability issues. Although this is not directly related to tax policy, if people are aware of the purpose of the policies, there is more likely to be a buy-in from the public. Westin (2005) states that environmental taxes lack a "voice," suggesting an institute dedicated to developing and advancing environmental taxation. Such an institute could help achieve acceptance of sustainable taxes and help form a comprehensive global policy.

Second, the policy should promote efficient use of and conservation of energy, water, and other resources. Elements of this portion of the policy could include incentives for the use of conservation measures, construction of energy-efficient buildings and machinery, and the use of renewable energy resources. Likewise, the policy could contain penalties for nonsustainable use of such natural resources.

Third, the policy should encourage the minimization of solid waste production. This could include incentives to implement the three "R's"—reduce, reuse, and recycle. Closely related is the fourth consideration, that of minimizing hazardous waste and toxic materials. This goal would include tax policies that focus on minimizing the use of toxic materials as well as strict penalties for failure to use, account for, and dispose of such waste responsibly.

Finally, the policy should provide tax incentives to encourage incorporation of sustainable design and planning principles in development, construction, and operation of infrastructure, grounds, and building. In addition to the more obvious tenet of designing sustainability into buildings, sustainable landscaping practices could be included. Additionally, planning could include a commitment to pedestrian travel, bicycle use, and other modes of transportation that promote a sustainable environment. This focus on transportation can include tax incentives for the purchase and use of bicycles, implementing environmentally friendly transit, and making the use of theses modes of transportation convenient to the public.

Coordination

As has been observed, we live in a global society. From a sustainable tax view, coordination has two implications. Companies that are not environmentally responsible may seek to relocate to an area with fewer environmental restrictions. Additionally, those areas with fewer environmental restrictions do not exist in isolation. Nonsustainable activities carried on in these areas will have a spillover effect, creating environmental and other difficulties that extend beyond their borders. A global sustainability tax agreement, coupled with other global sustainability agreements, is the most effective manner to isolate and change the behavior of noncooperating nations. When all major industrialized nations have ratified the agreement, the opportunity to shop for a "better" venue will be eliminated or greatly reduced. Such agreements could include provisions for tariffs on exported goods produced through nonsustainable processes in nonparticipating nations.

The Pigovian Approach

Pigovian taxes are designed to correct negative externalities that arise in the marketplace. There is no question that any number of negative sustainable actions occur in an open economy. Often, these externalities arise not from malice, but from ignorance or lack of the availability of a sustainable alternative. The issue of plastic bags is a prime example. Consumers have used these bags by the billions, primarily due to the lack of any incentive to seek alternatives. Once other alternatives became available, and the consumer was made aware of the problems created by plastic bags, their use declined. However, their use did not drop to levels most would consider acceptable. Therefore, a Pigovian solution was called for. When governments levied taxes on the use of plastic bags, their use declined significantly. Businesses were caught between the issue of paying the bag tax themselves, or passing it on to the consumer. Neither was seen as a workable solution, so alternatives to plastic bags were made available.

Removal of Subsidies

In somewhat of a "reverse Pigovian" approach, there are many tax subsidies in place that damage the environment and hamper sustainability efforts.

These subsidies should be eliminated. Among the culprits in this area are tax preferences for oil, mining, and timber. In the United States, a sport utility vehicle is eligible for tax breaks not available for passenger vehicles weighing under 6,000 pounds. The mortgage interest deduction has even been criticized for subsidizing home ownership and making second and larger homes more affordable. Removal of such subsidies and adoption of the Pigovian approach would have the effect of requiring polluters to pay taxes on their activities that are not environmentally friendly.

Social Equity

Social equity is another difficult issue in relation to sustainability. Any public policy will affect some members of society more than others. Steps must be taken to assure that the burden of sustainable taxation does not fall unjustly on low-income households. Government must ensure that sustainability tax policies do not lead the poor to pay a larger share of their income than wealthier households pay. This goal can be sought by paying a lump-sum to households affected by sustainability taxes. Other approaches would apply different rate structures on the basis of household income or exempt some groups from the tax measure (OECD 2006). A second aspect of social equity is dealing with nations that are poverty-stricken. Aid from industrialized nations can assist these countries in improving their economy as they undertake sustainability-oriented measures.

Visibility

A tax that is not visible and understood will not achieve a high level of support from the public, from companies, from leaders, and from nations. To make sustainable taxes visible and understandable, they should be distinct, nondiscriminatory, and defensibly quantified. A tax is distinct when the basis for setting the tax is clear and it is distinguished from other taxes. A nondiscriminatory tax should be applied to all similar sources of environmental and social damage. For example, coal, heating oil, and gas should all bear their share of the environmental tax, as each is a source of carbon dioxide and other pollutants. A tax is defensibly quantified if the proceeds from the tax are utilized to combat environmentally harmful activities rather than being viewed as a revenue measure (Newbery 2001).

Taxes in the Mix of Sustainable Policy Considerations

Taxes are not the only policy instrument in the hands of government for achieving a sustainable future. Indeed, they are only a part of the macroview of sustainability. Therefore, sustainable tax policy must be a part of an effective and efficient instrument mix. To achieve this goal, three requirements must be met.

First, there must be a good understanding of the environmental issue being addressed. Over 2,500 years ago, Chinese General Sun Tzu stated, "Know thyself, know thy enemy." Effective action comes when participants understand the environmental issues involved and why they are issues. Second, there must be a good understanding of how tax policy links with other policy areas. An effective policy will not be achieved if each policy area does not interact and coordinate effectively with the others. Finally, there must be a good understanding of the interactions between the different instruments in the mix. These instruments must not counterbalance each other (OECD 2007).

Conclusion

Taxes can be an effective tool in the policy mix to achieve sustainability on our planet. However, to be effective, the policy must be global in nature. A well-designed policy instrument similar in nature to the Kyoto Protocol will be the best hope for achieving this objective.

References

AECA. 2008. CO_2 Tax on cars widespread in West-European member states (March 24). http://www.acea.be/index.php/news/news_detail/co2_tax_on_cars_widespread_in_west_european_member_states. (Accessed January 20, 2010).

Barde, J. 2005. Implementing green tax reforms in OECD countries: Progress and barriers. In *Critical issues in environmental taxation,* ed. H. Ashiabor, K. Deketelaere, L. Kreiser, and J. Milne. Oxford: Oxford University Press.

Brechling, V., D. Helm, and S. Smith. 1991. Domestic energy conservation: Environmental objective and market failures. In *Economic policy toward the environment,* ed. D. Helm. Oxford: Blackwell, 263–264.

Carlson, W. 2008. Westport first in state to ban plastic bags. *New York Times,* September 26, 2008. http://www.nytimes.com/2008/09/28/nyregion/connecticut/28bagsct.html. (Accessed April 19, 2010).

Cassidy, J. 2009. An economist's invisible hand. *Wall Street Journal,* November 28, 2009. http://online.wsj.com/article/SB10001424052748704204304574545671352424680.html. (Accessed April 18, 2010).

European Commission. 2000. *European Economic Review, No. 71,* Chapter 4, Economic growth and environmental sustainability—A European perspective, European Commission, Brussels, 155–184.

Fischlowitz-Roberts, B. 2002. *Restructuring taxes to protect the environment.* Washington, D.C.: Earth Policy Institute.

Fujiwara, N., J. N. Ferrer, and C. Engehofer. 2006. The political economy of environmental taxation in European countries, CEPS Working Document #245. Brussels: Centre for European Policy Studies.

Furman, J. February 10, 2010. The concept of neutrality in tax policy. Brookings Institute.

Gale, W. G. 2009. The case for environmental taxes. www.brookings.edu/opinions. July 21, 2005. Accessed January 23, 2009.

HM Government. 2008. Building a low-carbon recovery. http://www.hm-treasury.gov.uk/d/Budget2009/bud09_chapter7_193.pdf. (Accessed January 23, 2010).

Lee, J. 2009. Taxing plastic bags, from pennies here to millions there. *New York Times,* February 2, 2009.

Marcoux, C. 2006. Explaining the U.S. decision to sign the Kyoto Protocol. Presented at the Annual Meeting of the International Studies Association, San Diego, CA, March 22, 2006.

McHugh, J. P. 2004. Adam Smith's principles of a proper tax system, Midland, MI: Mackinac Center for Public Policy. http://www.mackinac.org/6495. (Accessed January 19, 2010).

Milne, J. (Lead Author), N. Golubiewski, and C. J. Cleveland (Topic Editors). 2007. Environmental taxation in Europe and the United States. In: *Encyclopedia of earth*, ed. C. J. Cleveland (Washington, D.C.: Environmental Information Coalition, National Council for Science and the Environment). (Accessed January 3, 2010).

Morgenstern, R. 1995. Environmental taxes—dead or alive? Discussion paper dp-96-03, Washington, D.C.: Resources for the Future, 1995.

Newbery, D. M. 2001. Harmonizing energy taxes in the EU. Paper presented at the Tax Policy in the EU Conference. Rotterdam: Erasmus University.

OECD. 2001. *Environmentally related taxes: Issues and strategies*, Paris: OECD. http://oecd.org/publications/Pol_brief/. (Accessed February 6, 2010).

———. June, 2006. *The social dimension of environmental policy*. Paris: OECD. http://www.oecd.org/dataoecd/31/0/36958774.pdf. (Accessed February 6, 2010).

———. June, 2007. *The political economy of environmentally related taxes*. Paris: OECD. http://www.oecd.org/dataoecd/26/39/38046899.pdf. (Accessed February 6, 2010).

Owen, D. 2009. How traffic jams help the environment. *Wall Street Journal*, October 9, 2009. http://online.wsj.com/article/SB10001424052748703746604574461572304842840.html. (Accessed April 18, 2010).

Reusablebags.com. http://www.reusablebags.com/facts.php?id = 4. (Accessed February 6, 2010).

Roach, J. 2008. Plastic-bag bans gaining momentum around the world. *National Geographic News*, April 4, 2008. http://news.nationalgeographic.com/news/2008/04/080404-plastic-bags.html. (Accessed April 18, 2010).

Rosenthal, E. 2008. Motivated by a tax, Irish spurn plastic bags. *New York Times,* February 2, 2008. http://www.nytimes.com/2008/02/02/world/europe/02bags.html. (Accessed April 18, 2010).

Science Daily. 2008. Governments must explain benefits of environmental taxes, experts urge. http://www.sciencedaily.com/releases/2008/11/081128110715.htm. (Accessed January 23, 2009).

Sprinkel, O. 2009. What's your definition of sustainability? *Sustainable Brands Blog of Oliva Sprinkel* (April 8). http://sustainablebrands.wordpress.com/2009/04/08/whats-your-*definition-of-sustainability/*. (Accessed January 30, 2010).

Tax Policy Center. 2008. Taxes and the environment. Washington, D. C.: Urban-Brookings Tax Policy Center. www.taxpolicy center.org/briefingbook. (Accessed January 30, 2010).

Transportation alternatives. *Congestion pricing*. New York: Transportation Alternatives. http://www.transalt.org/campaigns/congestion. (Accessed February 6, 2010).

Tutor2u. 2009. Market failure—environmental taxation. http://tutor2you.net/economics/content/topics/marketfail/environmental_taxation.html. (Accessed January 23, 2009).

Ventosa, I. P. 2008. *Taxation, innovation and the environment—the Spanish case*. Paris: Organisation for Economic Co-operation and Development.

Watson, R. S. 2009. Harvesting tax benefits of green building incentives. *Journal of Accountancy,* August 2009. http://www.journalofaccountancy.com/Issues/2009/Aug/20091640.htm. (Accessed April 18, 2010).

Westin, R. 2005. Promoting environmental taxes. In *Critical issues,* ed. Deketelaere, Kreiser, and Milne, 2005.

PART III

Business Transformations of and by Its Environment

CHAPTER 10

Why Changing the Way We Use Energy Is Essential for Global Sustainability

Richard H. Jones

The founding mission of the International Energy Agency (IEA) was oil security. In the early 1970s, oil reigned supreme in the global energy sector. The 24 industrialized member countries of the Organization for Economic Co-operation and Development (OECD) accounted for 72 percent of world oil demand. The Organization for Petroleum Exporting Countries (OPEC) produced 50 percent of the world's oil. After key producers cut supplies to major consumer countries

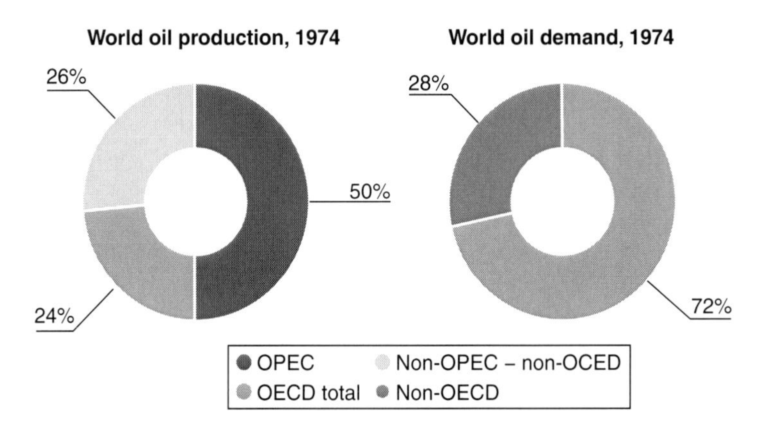

Figure 10.1 World oil production and demand in 1974

in the fall of 1973 (leading to what is commonly known as the "first oil shock"), the IEA was created in 1974.

The 15 founding members of the IEA (Austria, Belgium, Canada, Denmark, Germany, Ireland, Italy, Japan, Luxembourg, Netherlands, Spain, Sweden, Switzerland, the United Kingdom, and the United States) and Norway, which signed a separate agreement, had two key goals:

- To secure access to reliable and ample supplies of oil; and
- To establish and maintain effective emergency response capabilities in case of a supply disruption

Under the 1974 International Energy Program, the agreement establishing the Agency, IEA member countries agreed to hold stocks equivalent to at least 90 days of the prior year's net oil imports. In the event of a supply disruption, member countries would release stocks, restrain demand, or increase supply to restore stability to the oil market. Collective action would be taken only when physical oil supplies were affected, not in response to price movements. The IEA emergency response system has been fully activated twice—after the Iraqi invasion of Kuwait leading to the 1991 Gulf War and after hurricanes damaged oil production infrastructure in the Gulf of Mexico in 2005. In both cases, the system worked. The IEA response quickly offset concerns about supply shortfalls and stabilized global oil markets. On other occasions, mere knowledge that consultations on the use of stocks were under way has helped calm markets.

Beyond Oil to Climate Change

More than 35 years after its creation, the IEA continues to work to ensure that its now 28 member countries1 have reliable access to global oil supplies. Yet the Agency's role and mission have evolved, as have its member countries' energy-related policy objectives. New challenges have emerged. Energy security concerns are no longer focused solely on oil, but now extend across the energy sector to include assured access to all forms of energy, including natural gas and electricity. The dynamics among market players have changed too, as the more industrialized nations are no longer the major consumers of energy. In 2005, non-OECD countries collectively overtook the OECD as the largest consumers of energy. This trend is expected to continue. Just over 90 percent of the increase in world primary energy demand between 2007 and 2030 is projected to come from non-OECD countries. As a result, their share in world energy demand will grow from 52 percent to 63 percent. China

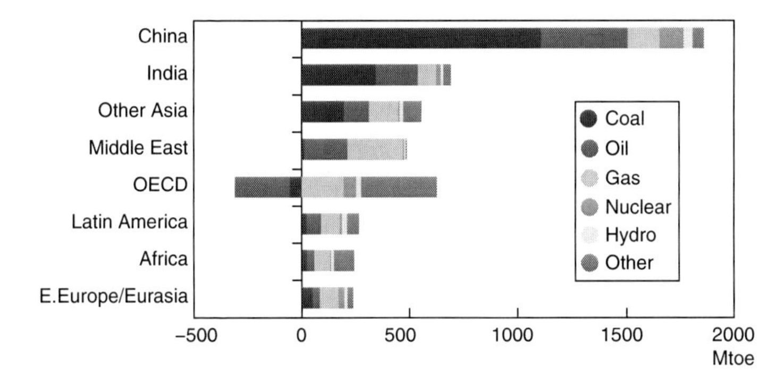

Figure 10.2 Incremental primary energy demand by fuel and region in the Reference Scenario, 2007–2030

and India are the main drivers of non-OECD demand growth, followed by the Middle East.

Another challenge that has become more urgent over time is reducing the environmental impact of energy production and use. While air pollution, deforestation, and habitat loss have all been linked to energy, its unquestioned role as one of the largest sources of greenhouse gas (GHG) emissions, especially carbon dioxide (CO_2), makes the energy sector one of the key contributing factors to climate change. Climate scientists believe that increased concentrations of GHGs in the atmosphere are responsible for rising global temperatures. The energy sector accounts for 84 percent of global CO_2 emissions and 64 percent of the world's GHG emissions. Maintaining current policies and inefficient uses of energy would rapidly increase our dependence on fossil fuels, leading to significant global warming and irreversible climate change. In light of the pivotal role of energy in meeting the climate challenge, IEA member countries, as well as the Group of Eight (G8) and Major Economies Forum (MEF), have tasked the Agency to conduct analyses and provide recommendations on what measures governments must take to drive a transition to cleaner, more competitive, and more efficient energy use. This work cuts across IEA activities involving economic analyses, technology roadmaps, energy efficiency recommendations, sectoral assessments, and, fundamentally, the collection of good-quality data. Any proposed policies to mitigate climate change must incorporate the "Three Es" of sound and sustainable energy policy: energy security, economic development, and environmental sustainability. IEA's analysis shows that effective solutions will also require that all nations—not only the member countries of the IEA—must participate in developing and implementing the new policies and actions that will be required.

The World Energy Outlook Assessment

Growing Urgency to Act to Mitigate Climate Change

Climate change is one of the greatest challenges facing the world today. There is an increasing global consensus around the science, which tells us that anthropogenic emissions of GHGs, including CO_2 are causing the planet to warm and climate patterns to change; and that abatement of GHG emissions is therefore necessary. Due to the ease with which GHGs can circulate within the atmosphere, this abatement cannot be achieved by any one country acting alone. The scale of the associated change means that action must be global.

A certain amount of climate change and global warming have already taken place, and more is already locked-in. Atmospheric concentrations of GHGs are already more than double what they were in preindustrial times, and the long life of CO_2 in particular guarantees that, even if all emissions were halted today, some further warming would take place before carbon sinks could absorb the excess CO_2 remaining in the atmosphere. There is also uncertainty around the point at which feedback loops will kick-in, meaning that climate change becomes a vicious circle, as warming leads to greater natural emissions, leading to further warming, and so on. Delaying action will only increase costs as well as risks.

Consequences of Staying on the Current Path

Energy-related CO2 is the major component in global emissions of GHGs. In 2007, total energy-related emissions were 28.8 gigatons (Gt). According to the IEA Reference Scenario (i.e., a "no new policy" scenario) analysis, if the world stays on its current path, energy-related emissions will reach 34.5 Gt in 2020 and 40.2 Gt in 2030. In this scenario, emissions do not rise in all parts of the world. OECD emissions would be 3 percent lower in 2030 than in 2007. By contrast, all major non-OECD countries would see their emissions rise. Of the 11 Gt growth in global emissions, China would account for 6 Gt, India for 2 Gt, and the Middle East for 1 Gt.

All sectors would see growth in energy-related CO_2 emissions, with aviation and power generation being the fastest growing sectors. The power sector accounts for over half the increase in emissions between 2007 and 2030, with a 60 percent increase from coal-fired generation.

This rapid growth of global GHG emissions would lead to a substantial long-term increase in the concentration of GHGs in the atmosphere and a resulting large increase in global temperature. IEA's projections for energy-related CO_2 emissions from today to 2030 lie within the range modelled by the Intergovernmental Panel on Climate Change (IPCC). Taking into account emissions of all GHGs from all sources, the Reference Scenario corresponds to a long-term atmospheric concentration of GHGs equivalent to around 1,000 parts per million (ppm) of CO_2 equivalent.

Climatic Consequences

The 1,000-ppm trajectory is estimated by the IPCC to yield a global mean temperature rise of around 6°C by the end of this century. At this level, studies indicate that the environmental impacts would be severe. They include sea-level rise of up to 3.7 meters, causing the loss of several islands, 50 percent loss of coastal wetlands, droughts, floods, destruction of habitats, species extinctions, decreased production of cereal crops, and food shortages. There is also a high risk of dangerous feedbacks and an irreversible cycle of environmental destruction.

Achieving the Necessary Abatement: The 450 Scenario

Although opinion is mixed on what might be considered a sustainable long-term level of annual CO_2 emissions for the energy sector, the need to limit the global temperature increase to no more than 2°C was agreed upon in the "Copenhagen Accord" reached at the December 2009 UN climate talks. To limit the probability of a global average temperature increase in excess of 2°C to 50 percent, the concentration of GHGs in the atmosphere would need to be stabilized at a level equivalent to around 450 ppm of CO_2. To put the world onto this 450-ppm trajectory, we need a veritable low-carbon revolution, a wholesale and rapid transformation in the way we produce, transport, and use energy. Though difficult, our analysis suggests that such a revolution can be achieved with international cooperation and the right policy framework.

First, energy needs to be used much more efficiently and the carbon content of the energy we consume must be reduced, largely by switching to low- or zero-carbon sources. The IEA flagship publication *World Energy Outlook (WEO)* 2009 shows how this objective can be achieved in the so-called 450 Scenario, through farsighted and coordinated policy action across all regions. In this scenario, global energy-related CO_2 emissions peak at 30.9 Gt just before 2020 and decline thereafter to 26.4 Gt in 2030—2.4 Gt below the 2007 level and 13.8 Gt below that in the Reference Scenario. These reductions result from a plausible combination of policy instruments, notably steps to encourage energy efficiency, carbon markets, sectoral agreements, and national policies and measures tailored to the circumstances of specific sectors and groups of countries. Only by taking advantage of mitigation potential in all sectors and regions can the necessary emission reductions be achieved. "OECD + " countries—a group that includes the OECD and non-OECD European Union countries—are assumed to take on national emissions-reduction commitments by 2013. The 450 Scenario assumes that all other countries will adopt appropriate domestic policies and measures, including generating and selling emissions credits. After 2020, commitments would be extended to other major economies, which include China, Russia, Brazil, South Africa, and the Middle East.

In the 450 Scenario, primary energy demand grows by 20 percent between 2007 and 2030. This increase corresponds to an average annual growth rate of 0.8 percent, compared to 1.5 percent in the Reference Scenario. Increased energy efficiency in buildings and industry reduces the demand for electricity and, to a lesser extent, fossil fuels. The introduction of electric vehicles, hybrids, and more efficient internal combustion engines, as well as second-generation biofuels, will reduce average emissions intensity of new cars by more than half, cutting oil needs. Thus, the share of zero-carbon fuels in the overall primary energy mix increases from 18 percent to 32 percent in 2030, when CO_2 emissions per unit of gross domestic product (GDP) are less than half their 2007 level. Yet, continued economic growth means that global demand for all fuels, except coal, will be higher in 2030 than in 2007, and fossil fuels will remain the dominant energy sources in 2030.

End-use efficiency is the largest contributor to CO_2 emissions abatement in 2030, accounting for more than half of total savings compared with the Reference Scenario. Energy-efficiency investments in buildings, industry, and transport usually have short payback periods and negative net abatement costs, as the fuel-cost savings over the lifetime of the capital stock often outweigh the additional capital cost of the efficiency measure, even when future savings are discounted.

Decarbonization of the power sector also plays a central role in reducing emissions. Power generation accounts for more than two-thirds of the savings in the 450 Scenario (of which 40 percent results from lower electricity demand). There is a big shift in the mix of fuels and technologies in power generation: coal-based generation is reduced by half, compared with the Reference Scenario in 2030, while nuclear power and renewable sources make much bigger contributions. The United States and China together contribute about half of the reduction in global power-sector emissions. The emerging technology of carbon capture and storage (CCS) will be required to play a key role in the power sector (and in industry), and would represent 10 percent of total emissions savings in 2030, relative to the Reference Scenario.

The transport sector also needs to be decarbonized. Measures in the transport sector to improve fuel economy, expand biofuels, and promote the uptake of new vehicle technologies—notably hybrid and electric vehicles—lead to a big reduction in oil demand. By 2030, transport demand for oil is cut by 11 mb/d (million barrels per day), equal to more than 70 percent of all the oil savings in the 450 Scenario. Road transport accounts for the vast majority of these transport-related oil savings. A dramatic shift in car sales occurs; by 2030, conventional internal combustion engines represent only some 40 percent of sales (down from more than 90 percent in the Reference Scenario) as hybrids take up 30 percent of sales, and plug-in hybrids and electric vehicles account for the remainder. Efficiency improvements in new aircraft and aviation biofuels save 1.5 mb/d by 2030.

Costs of Moving to a Low-Carbon Economy

Compared to the Reference Scenario, the 450 Scenario entails $10 trillion more invested in energy infrastructure and energy-related capital stock globally during the projection period. Around 47 percent of incremental investment needs, or $4.7 trillion, are in transport. Additional investment (which includes the purchase of energy-related equipment by households in this analysis) amounts to $2.5 trillion in buildings (including domestic and commercial equipment and appliances), $1.7 trillion in power plants, $1.1 trillion in industry, and $0.4 trillion in biofuels production (mostly second-generation technologies, which become more widespread after 2020). More than three-quarters of the total, which is geographically distributed almost equally between OECD + countries and the rest of the world, is needed in the 2020s. On an annual basis, global additional investment needs reach $427 billion (0.51 percent of GDP) in 2020 and $1.2 trillion (1.1 percent of GDP) in 2030. Most of this investment would need to be made by the private sector; households alone are responsible for around 40 percent of the additional investments in the 450 Scenario, with most of their extra expenditure going to low-carbon vehicle purchases. In the short term, the maintenance of government stimulus efforts is crucial to this investment.

Taking into account financial flows between countries, the rebalancing of supply and demand, and the resulting changes in price levels, the 450 Scenario estimates that global GDP would be reduced in 2020 by between 0.1 percent to 0.2 percent, and in 2030 by between 0.9 percent and 1.6 percent compared with the Reference Scenario. Although these costs would be likely to fall more heavily in the later part of the period and could be substantial in individual years, they would be far lower than the costs imposed by the climate consequences of higher stabilization trajectories. Further, as the global economy is assumed to double between 2007 and 2030, a 1.6 percent fall in GDP in 2030 is equivalent to losing a few months of growth over 23 years. To put this impact into perspective, the 2008/2009 recession cost world growth as much in a single year.

Benefits of Moving to a Low-Carbon Economy

The cost of the additional investments needed to put the world onto a 450-ppm path would be offset by economic, health, and energy-security benefits. Energy bills in transport, buildings, and industry are reduced by $8.6 trillion globally over the period 2010–2030 and would continue after that time. Fuel-cost savings in the transport sector alone amount to $6.2 trillion over the projection period. Oil and gas imports and their associated bills in the OECD and developing Asia are much lower than in the Reference Scenario; in OECD countries they are lower than they were in 2008. Other implications include a large reduction in emissions of air pollutants, particularly in China and India, and in the cost of installing pollution-control equipment. Lower mortality rates could lead to as much as a billion man-years of increased longevity.

Financing the Transition

It is widely agreed that developed countries must provide more financial support to developing countries in reducing their emissions, but the level of support, the mechanisms for providing it, and the relative burden across countries are matters for negotiation. There is a wide range of potential funding approaches. In the 450 Scenario, by 2020 $197 billion in additional annual investment is required in non-OECD countries. The Copenhagen Accord committed to $100 billion in annual international investment to finance mitigation and adaptation in developing countries.

There are various channels through which funds can flow to developing countries. Carbon markets will undoubtedly play an important role. Depending on how international carbon markets evolve, primary trading of CO_2 emissions reductions between OECD + and other regions could range between 0.5 Gt and 1.7 Gt per year by 2020. A central case for a single international market sees a CO_2 price of around $30 per ton and annual primary trading of around $40 billion. The current Clean Development Mechanism (CDM) would need extensive reform to cope efficiently and robustly with a substantially increased level of activity. International funding pools, such as the Copenhagen Green Climate Fund mentioned in the

Accord, are other important channels that could provide a means of increasing financial transfers to developing countries.

The Cost of Delayed Action

The costs of delaying action are quite significant. A given climate change goal can be looked at in terms of a global "budget" of emissions over a given period of time, and which cannot be surpassed. A budget of 1 Tt CO_2 for the period 2000–2049 would give us a 75 percent chance of limiting temperature increase to 2°C, while a budget of 1.44 Tt CO_2 is broadly consistent with the 450 Scenario. A budget of 1.55 Tt between 2000 and 2049 is consistent with stabilization at 550 ppm and the much more severe implications of such a CO_2 level. The cumulative CO_2 emissions associated with this year's Reference Scenario are 2.1 Tt (Meinshausen et al. 2009). We are currently eating into these CO_2 budgets at a disproportionate rate (see figure below). Between 2000 and 2009, the world emitted a total of 313 Gt of CO_2—or some 31 percent of the budget of 1 Tt for the period. The Reference Scenario sees cumulative emissions since 2000 pass the 1 Tt level as early as 2028 and by 2049, they exceed 2 Tt. Even in the 450 Scenario, cumulative emissions to 2030 are substantially above the level that would distribute an emissions budget of 1.44 Tt evenly over time. And, it will be recalled that even this 1.44 Tt CO_2 budget still carries with it a 50 percent probability of a global temperature increase greater than 2°C. These estimates underline the urgency of prompt action to get back on track.

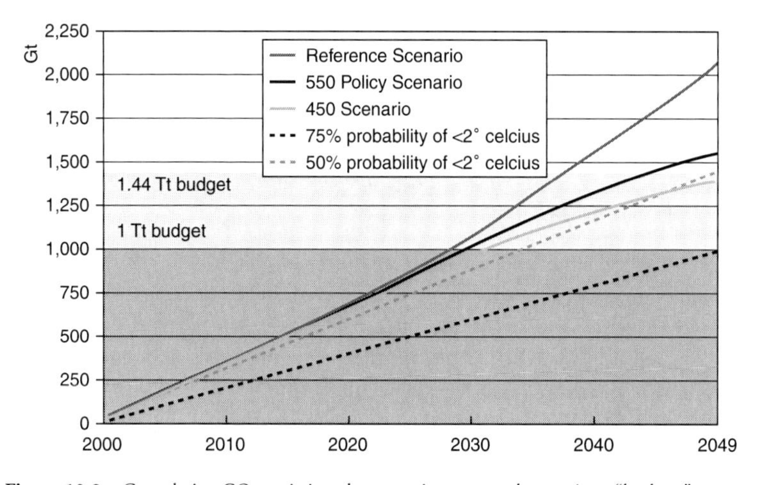

Figure 10.3 Cumulative CO_2 emissions by scenario compared to various "budgets"

This "budget analysis" highlights one fundamental fact: the opportunities to stabilize levels for the concentration of GHGs in the atmosphere at acceptable levels are diminishing rapidly. Emissions in the first decade of this century have probably already rendered a 75 percent probability of limiting temperature rise to 2°C out of reach, though a 450 Scenario is still achievable with urgent action. Delay carries the ever increasing costs needed to achieve even tighter annual emission levels in the future or suffering the consequences of still higher emission levels if action is not taken.

If the world decides to pursue a 450-ppm trajectory, every year of delay in implementing appropriate policies means subsequently catching up on abatement, with the risk of increased costs for further mitigation efforts. An indicative guide, based on the IEA results, is that for every year of delay before moving to a 450-ppm trajectory, an extra $500 billion is added to the global bill of $10 trillion for mitigating climate change. This figure applies only to delays of one–three years; further delay means that a 450-ppm trajectory becomes unattainable and the additional adaptation costs would be several times this figure. This result is highly sensitive to assumptions about marginal abatement costs at different points in time, although it is broadly consistent with the results in the limited literature available (van Vliet, den Elzen, and van Vuuren 2009; Keppo and Rao 2007; and Richels et al. 2007).

The costs of delay are in large part due to the inertia inherent in the energy sector, reflecting the long life of costly capital assets. The issue of lock-in in the energy sector highlights the importance of ensuring that capital expenditure, whether to expand or to replace capacity, focuses on low-carbon technologies so that it is these that become locked into the system. For every year that passes, the window for action on emissions over a given period becomes narrower—and the costs of transforming the energy sector to deliver a 450 Scenario increase.

Where We Stand Currently—Outcome of Copenhagen and How Far It Takes Us

In December 2009, the 15th Conference of the Parties (COP) to the United Nations Framework Convention on Climate Change (UNFCCC) resulted in the Copenhagen Accord. This was not, as had been hoped, a legally binding framework of targets and mechanisms for addressing climate change, but instead a high-level agreement on the goal of climate change policy—that warming be limited to no more than 2°C—and on some aspects of finance. Countries agreeing to the accord were asked to enter into an annex their abatement targets or the actions that they would take to mitigate climate change. Of the 136 countries that have indicated agreement to the accord, so far 82 have entered target or policy measures that they intend to take.

These targets and promises of action represent significant progress, as does the new international consensus around both the ultimate aim of climate change and the need to mobilize significant assistance in order to finance the dramatic transition to a low-carbon economy outlined above. However, these pledges do not go nearly far enough. The implementation of the most stringent versions of the current national pledges announced by a number of countries would bring global energy-related CO_2 emissions in 2020 to 31.7 Gt.

If policies were put in place in OECD and non-OECD countries to stabilize global emissions at this 31.7 Gt level from 2020 to 2030 (a substantial departure from the 40 Gt emitted in the Reference Scenario), the world could be on track for a 550-ppm stabilization trajectory. This scenario would be a major improvement on the Reference Scenario, in terms of its environmental consequences, but would still yield a temperature rise of around 3°C by 2100. This level of temperature increase would entail significant adaptation costs, as the global community came to terms with the rise in sea levels from melting ice caps and with a considerable increase in arid land in many parts of the world from changing weather patterns.

Current targets therefore represent a good start, but do not go far enough. If climate change is to be limited to the Copenhagen Accord's goal of a maximum 2°C increase in average temperatures, and if the cost of achieving this target is to be minimized, further action must be taken and must be taken promptly.

The Challenge of Communication

The implications of the analysis of the WEO 2009 are worrisome, to say the least. Our current path not only leads to higher concentrations of GHGs in the atmosphere and accelerated climate change, but also results in greater dependence on imported fossil fuels, which undermines energy security. In addition, the longer we wait to act, the more difficult and costly the solution becomes. That message is clear. The question is: Why are we waiting? Perhaps the answer is uncertainty—uncertainty about the costs and benefits of changing a system that has served much of the world well in many ways for more than a century.

To overcome such uncertainty, effective communication is critical. Policymakers, stakeholders, industry, and the public need to be better informed. The options for response must be more clearly identified.

Trade-offs must be carefully weighed and explained. The costs of action—and nonaction—must be assessed and compared. And the implications for daily life must be made clear. Providing such information sounds like a major task, and it is.

Both energy security and climate change are highly technical topics that can be quite daunting to a nonexpert. It takes time to understand what the construction of a 150-megawatt (MW) power plant means in terms of additional energy supply, why the concentration of carbon at 450 ppm is relevant, or how carbon markets function. And once understood, the message is unsettling—we must change the way we produce and use energy now or we face increased dependence on fossil fuels and catastrophic climate change. While some question this conclusion, others feel overwhelmed because the challenges seem so many and so enormous that any single change in behavior seems certain to have no effect. And the "free rider" problem surfaces—Why should one person (or company or government) sacrifice, incur costs, and change behavior for the greater good when their neighbour may not do the same?

In terms of overall priorities, both individuals and governments have immediate pressing concerns: employment, education, health care, etc.—especially during the current financial crisis and global recession. To many, energy security and climate change may seem abstract and remote, not directly and immediately affecting their wallets, careers, or health. And if they are perceived as concerns, then they are concerns to be dealt with in the longer term when more pressing problems have been handled. This stance is often justified through assertions that markets will automatically adjust to reduce imbalances and technology will find the solution to all problems.

In short, many excuses, none of them very compelling, exist for not strengthening energy security and mitigating climate change. The need to reduce these uncertainties and confusions about the global energy situation has led the IEA into expanding its communication commitments and activities.

Sending the Right Message to the Right Audience

The traditional audience for IEA analyses has been government officials and policymakers in member countries. The Agency and its work are well-known in the energy sector and related government ministries across consuming and producing countries. IEA analyses and recommendations have contributed to and, in some cases, helped guide policy formulation. In tackling global challenges like energy security and climate change, however, action needs to happen on a global scale, at levels that include other stakeholders, industry, and the public. Everyone needs to be involved.

In terms of reaching audiences beyond the IEA membership, the Agency is undertaking growing efforts to work with key producer and consumer countries, especially China, India, and Russia, and to include close cooperation with them on topics ranging from energy efficiency to low-carbon technologies. Ministers from all three countries participated in the 2009 IEA Ministerial meeting, and agreements were signed to further joint work. Numerous workshops have been planned and a number of events, including presentations highlighting IEA findings, have been organized. The IEA attributes a key role to communication and the media and is targeting journalists worldwide. Today, the Agency's work is regularly covered in the international press, both in IEA member countries and beyond, helping familiarize a broader public with IEA messages. Several recent IEA publications have been translated into both Chinese and Russian, and web pages in both languages are maintained on the IEA Web site. A new IEA training and capacity-building program is being established to formalize ongoing training for statisticians and other government energy officials from these and other nonmember countries. Other regions in which the Agency is most active include the Middle East, the Association of Southeast Asian Nations (ASEAN), and Latin America.

In addition, the IEA is working to consolidate and focus its ties with industry and the private sector. For years, the Agency has interacted with business groups and individual companies involved in the energy sector. From the Coal Industry Advisory Board to company secondments at IEA headquarters in Paris, to wide industry participation in the IEA Energy Technology Network (Implementing Agreements), these ties have been ongoing and constant. The more recent Chief Technology Officers' roundtable and Energy Business Council convene high-level discussions on energy-related challenges, industry concerns, and policy solutions. Further coordination efforts will enable the Agency to engage more effectively with non-IEA member countries and to take an increasingly proactive approach to energy-related public and private sector challenges. These efforts include close cooperation with the World Business Council for Sustainable Development (WBCSD), the creation of new initiatives to accelerate the spread of low-carbon energy technology, and the development of a partnership program for energy security and sustainability.

Capturing the attention of the public at a time when too much information is available is especially challenging. New types of communication technology offer opportunities but require time, maintenance, and targeted communications. The IEA Web site has transformed from an archival source of information to an active entity that is updated daily and seeks to encourage visitors to return. More papers and documents are available for free, while links to related sites provide access to further information. Efforts to convey complex, technical messages for a broader audience are underway, including condensed, less technical versions of publications such as the WEO and the launch of a webpage on basic energy efficiency.

Some recent examples of initiatives to reach the public include a video on energy efficiency to demonstrate steps that individuals can take to cut their energy consumption on a daily basis. This video may be picked up by a major television network as a public service announcement. Other videos interviewing IEA experts have been posted to explain their work. A Facebook page was launched to attract new "friends" and interest them in IEA work and messages. The goal of all these efforts is to enable greater access to, and understanding of, the Agency's work and, more importantly, to get the messages about its work to new audiences.

Offering Solutions

The next step is to present the path forward. What can governments, industry, and the public do to tackle energy security concerns and mitigate climate change?

In recent years, the IEA's findings have focused on identifying specific policies and steps that governments may implement to achieve specific goals. The climate analysis in WEO 2009 included specific measures to hold CO_2 concentrations in the atmosphere to 450 ppm. The 25 IEA energy efficiency recommendations presented to the G8 and IEA member countries in recent years presented clear steps to cut energy consumption. The IEA's *Energy Technology Perspectives* series and its Technology Roadmaps show what specific mixes of energy technologies are required to meet broader policy goals. In addition, the Agency increasingly works to calculate and compare the costs of such measures—a key consideration for any government.

In terms of messages for the private sector, the WEO has long highlighted the importance of clear and consistent policies to ensure investment. IEA technology work increasingly looks at different sectors, including industry, chemicals, services, transport, and buildings. A 2009 roadmap focused on the cement sector. An IEA study of lighting and subsequent activities with industry helped bring the phase-out of incandescent light bulbs. Further work—conducted both with industry and governments—on best practices in appliances, buildings, and transport could all have long-term impacts on standards and other regulations.

And for the general public, IEA's goal is to present messages in a manner that is interesting, easy to understand, and relevant to daily life. We want people to recognize why they need to take steps to help increase energy security and counteract climate change, and how they can take those steps. And we are working to make it relatively simple and inexpensive for them to do so. Only then will we truly be on a path to a sustainable energy future.

Notes

1. IEA member countries: Australia, Austria, Belgium, Canada, Czech Republic, Denmark, Finland, France, Germany, Greece, Hungary, Ireland, Italy, Japan, Korea (Republic of), Luxembourg, Netherlands, New Zealand, Norway, Poland, Portugal, Slovak Republic, Spain, Sweden, Switzerland, Turkey, United Kingdom, United States. The European Commission also participates in the work of the IEA.

References

IEA: International Energy Agency. 2009. *World Energy Outlook 2009*. Paris: International Energy Agency.

Keppo, I., and S. Rao. 2007. International climate regimes: Effects of delayed participation. *Technological Forecasting and Social Change* 74 (7): 962–979.

Meinshausen, M., N. Meinshausen, W. Hare, S. C. B. Raper, K. Frieler, R. Knutti, D. J. Frame, and M. Allen. 2009. Greenhouse gas emission targets for limiting global warming to 2°C. *Nature* 458 (7242): 1158.

Richels, R., T. Rutherford, G. Blanford, and L. Clarke. 2007. Managing the transition to climate stabilization. Working Paper. AEI-Brookings Joint Center for Regulatory Studies. Washington, D.C.: AEI-Brookings Joint Center for Regulatory Studies.

Van Vliet, J., M. G. J. den Elzen, and D. P. van Vuuren. 2009. Meeting radiative forcing targets under delayed participation. *Energy Economics* 31, Supplement 2 (December): S152–S162.

CHAPTER 11

From Carbon to Carbohydrates: Toward Sustainability in the Plastics Industry

Avrath Chadha

Introduction

With a global polymer production of over 250 million tons in 2008, the plastics industry is one of the chemical industry's major sectors (*PlasticsEurope* 2009). Petro-based polymer technology has been the dominant production technology for plastics (Vellema, van Tuil, and Eggink 2003) and the industry has tended to endorse incremental process-oriented innovations rather than large-scale product or process innovations (Kaplan 1998). However, in the coming years, radical change in the plastics industry might be inevitable (Sartorius 2003). Since most polymers are produced from fossil-fuel raw materials, their depletion is a widely recognized issue in the industry (Mohanty, Misra, and Drzal 2002). Some studies estimate oil reserves will run dry in approximately 30–50 years (Busch and Hoffmann 2007; Energy Information Administration 2007; International Energy Agency 2007). However, the economic impact of oil depletion could hit the plastics industry much earlier, especially if rising oil prices lead to product cost increases that cannot be passed on to customers (Platt 2006; Ren 2003).

To gain independence from crude oil, the plastics industry has started to focus on the development of technologies using materials that substitute for petro-based polymers. In line with a "green chemistry" orientation and an emphasis on renewable resources, considerable research and development efforts are currently focusing on so-called biopolymers or bioplastic. These materials are processed from renewable feedstock like starch and may or may not yield end products that are biodegradable (*European Bioplastics* 2006; Mohanty, Misra, and Hinrichsen 2000).

Although market reports on biopolymer technology suggest that technological progress and economies of scale in production will lower costs and increase the potential for petro-based biopolymers to substitute for petro-based polymers in the years ahead (Crank and Patel 2005; Platt 2006), the extent of the industry-wide development of biopolymer technology is not well understood (Michael 2004; Paster, Pellegrino, and Carole 2003; Runge 2006). In addition, the market penetration of biopolymers depends not only on scientific progress but also on strategic management decisions of the involved firms. Some pioneering firms are already on the market while others are still waiting. This chapter applies a case study method to explore the different reasons firms have for entering the biopolymer market and analyzes the various corporate strategies they use for biopolymer technology development.

Biopolymers: Emerging Plastics for the Environment

Soaring oil prices, growing worldwide concern about greenhouse gas emissions, consumer interest in sustainability issues and renewable resources, and an emphasis on waste management are yielding increased investment in biopolymer technology. In the European Union, environmental concerns and legislative incentives are particularly effective in stimulating interest in biopolymers (Lee and Park 2003). New technologies in plant breeding and processing are narrowing the cost differential between biopolymer and petro-based polymer as well as improving material properties such as processing viscosity (Bastioli 2005; Paster, Pellegrino, and Carole 2003). Therefore, advances in biotechnology, especially in genetic engineering, will play an important role in the further development of biopolymers.

According to Dahlin's and Behrens' (2005) criteria, biopolymer technology resembles a radical innovation: because of its different feedstock material, biopolymers are significantly different from previous or current innovations in the industry such as process improvements on petro-based plastics or new products such as petro-based biodegradable polymers. Furthermore, they are likely to have an impact on future innovations in the industry as oil scarcity becomes more severe (Bastioli 2005; Mohanty, Misra, and Drzal 2002). Compared to petro-based polymers, biopolymers have a number of novel material characteristics, such as antistatic properties, good printability, and high degree of water permeability, which make them suitable for a wide range of applications in agriculture, automotive, medicine, and packaging (Beucker and Marscheider-Weidemann 2007; Crank and Patel 2005). Initial applications such as loose foils, bottles, trays, crockery, et cetera have already established themselves successfully in important markets (Paster, Pellegrino, and Carole 2003; Sartorius 2003). In addition to these short-lived applications, there are also durable applications, for example, injection molded parts from polylactic acid (PLA) that can now be found in products in the automotive and electronics industries (Bastioli 2005; Platt 2006).

Three classes of biopolymers can be distinguished (Mohanty, Misra, and Drzal 2002; Ren 2003; Sartorius 2003): (1) biodegradable polymers based on renewable resources, (2) biodegradable polymers based on petro-based resources, and (3) nonbiodegradable polymers based on renewable resources. Figure 11.1 shows the annual worldwide production capacity for these materials.

Currently, three different kinds of biopolymers have reached the commercialization stage and are available on the market: (1) starch materials, (2) PLA, and (3) cellulose materials (Bastioli 2005). To date, starch polymers represent by far the largest group of commercially available biopolymers. Simplistic products such as pure thermoplastic starch and starch/polyolefin blends were first introduced. Research and development currently also embraces nanotechnology, for example, nanoparticle starch fillers as a substitute for carbon black in tires. Major progress has also been made in the production of PLA. Several plants for the manufacture of PLA have been established in the recent years.

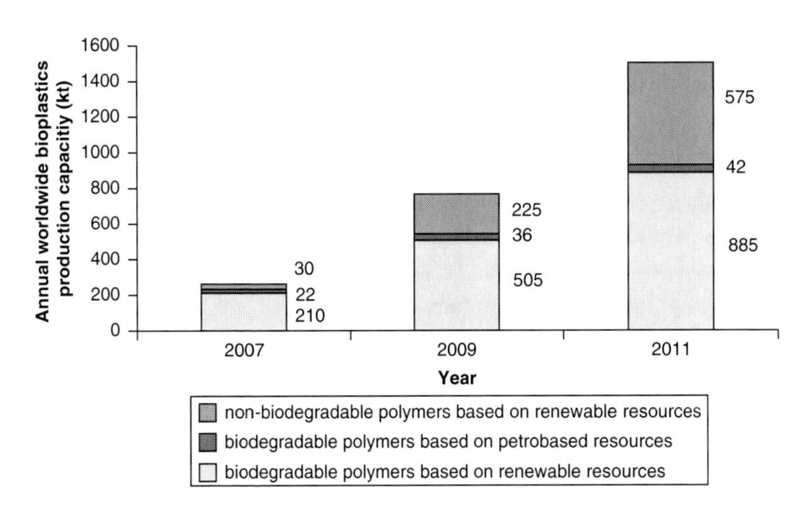

Figure 11.1 Annual worldwide bioplastics production capacity

Apart from being the monomer for PLA, lactic acid also has the potential to become a new bio-based bulk chemical from which a variety of other chemicals and polymers can be produced, for example, acrylic acid and propylene glycol. For a number of years, cellulose polymers have played a key role in a wide range of applications, such as those in apparel production. However, those biopolymers have lost important markets to polyolefins and other substitutes. Recently, attempts have been made to develop new cellulose polymers markets in the area of films and natural fibers composites (Crank and Patel 2005; Platt 2006).

A promising fourth class of biopolymers is so-called polyhydroxyalkanoates, which are natural polyesters synthesized from carbonaceous substrates by bacterial species. These biopolymers exhibit a broad range of mechanical properties from rigid plastics to ductile elastics. They are produced using fermentation processes of glucose and organic acid. However, polyhydroxyalkanoates are still commercially available only in very limited quantities (Platt 2006).

Methodology

The research design for this chapter is a multiple-case, inductive study of ten U.S., German, and Swiss companies active in the plastics industry. The cases selected met two requirements: first, firms had to be raw material suppliers or plastics converters, since such companies are most affected by the new biopolymer technology. This focus on raw material suppliers and converters means that the majority of the polymer value chain is included in the analysis. Second, firms had to be in the process of developing or had already developed biopolymer technology.

Data were collected from two main sources: archival data and semistructured interviews (Eisenhardt and Graebner 2007). Over a hundred public documents, including newspapers, press releases, market reports, trade publications, company accounts, and annual reports were collected and analyzed. Additionally, full-text queries on the Business Source Premier and LexisNexis databases were run using the names of the companies along with the keywords "biopolymer," "bioplastics," and 15 similar terms, for example, product names or names of chemical raw materials.

In total, 36 semistructured interviews were conducted. As shown in Table 11.1, the interviewees were representatives from different corporate functional units such as general management, sales, research and development, and sustainability.

This approach yielded a comprehensive overview of the development decision situation for biopolymer technology in the respondent companies.

Table 11.1 Characteristics of firms studied and interviewees

Firm	*Country*	*Industry*	*Number of Employees*	*Interview-p artners*
A	Germany	Raw Material Supply	100,000	General Management; Head of R&D; Head of Marketing; Head of Environmental Mgmt.
B	Switzerland	Raw Material Supply	100	Member of Management Board; Head of R&D; Head of Marketing; Head of Environmental Mgmt.
C	Germany	Packaging	2,500	Head of R&D; Head of Marketing
D	USA	Packaging	2,000	Head of Application Technology; Head of Marketing
E	Germany	Automotive	350,000	Head of R&D; Head of Application Technology; Head of Corporate Development; Head of Environmental Mgmt.
F	USA	Multi-Technology	80,000	General Management; Head of R&D; Head of Application Technology; Head of Marketing; Head of Environmental Mgmt.
G	USA	Raw Material Supply	15,000	Head of R&D; Head of Application Technology; Head of Marketing
H	Germany	Packaging	150	Member of Management Board; General Management; Head of R&D; Head of Corporate Development
I	Germany	Automotive	300,000	Divisional R&D Manager; Head of Application Technology; Divisional Marketing Manager; Head of Environmental Mgmt.
J	Switzerland	Automotive	1,000	Member of Management Board; Managing Director; Head of R&D; Head of Corporate Development

The interviews were approximately 60–90 minutes in length and started with a brief description of the interviewees' professional background and their role in the organization. General questions on fossil resource scarcity and its implication for the company and on the reasons for adopting biopolymer technology were asked to gain a better understanding of the decision situation the companies were in. Later in the interviews, the focus shifted to capabilities and competencies that were utilized to develop biopolymer technology. For instance, interviewees were requested to describe the innovation process and illustrate their specific role in it. The interviewees then explained the routines and methods they believe led to the development of biopolymer technology.

Each interview was taped and then subsequently converted into an interview protocol for use in the data analysis. Interview protocols were verified for accuracy by the interviewees. Follow-up questions were asked by phone or via e-mail when clarification was needed. For data analysis, the first step involved an in-depth examination of the initial interview data to uncover key themes, and then the individual case studies based on interview protocols and archival data were thoroughly scrutinized (Miles and Huberman 1994). Finally, a cross-case analysis was conducted to compare and analyze the data for competence patterns (Eisenhardt 1989).

Results

Reasons for the Increased Acceptance of Biopolymers

Raw material suppliers as well as converters confirmed that fossil-fuel scarcity and the resulting increased prices for crude oil are one of their major strategic challenges. Because of the strong competition from petro-based polymers in a highly price-driven market, increased costs for raw materials can only be passed on to the customer with great difficulty. All of the examined firms are trying to secure their fossil raw material base by developing specific sourcing strategies such as long-term supplier agreements. Parallel to this strategy, raw material suppliers are scanning the market for alternatives to petro-based polymers and are focusing their research efforts on biopolymers.

Interestingly, fossil-fuel scarcity is not the trigger for every firm to shift the attention toward biopolymers. Some firms are motivated by increasing queries from customers for more sustainable product solutions. For example, converters are confronted by a rising demand for biopolymers by brand owners and end consumers. Activities of competitors are another reason for raw material suppliers and converters to enter the still very young market for biopolymers. Raw material suppliers see biopolymers as a feature distinguishing their products from those of their competitors. Many such firms have added biopolymers to the petro-based polymers in their portfolio. Market and technology leadership is especially important for large raw material suppliers and firms active in the automobile sector and is a declared corporate goal actively communicated internally. All firms interviewed also view the "green image" of biopolymers as a very positive factor, although they find it difficult to quantify.

Corporate Strategies for Biopolymers

All studied firms were active in the market for petro-based polymers prior to the development of biopolymer technology. Over the years, petro-based polymer technology has achieved market dominance and firms in the sample have evolved to suit a specific set of requirements for petro-based polymers, such as the use of specific processing machinery or business processes necessary for production. These established processes resulted in inertia as many divisions and departments refrained from sharing knowledge or manpower as attention to biopolymers increased. Inertia often hindered proposed changes and innovations, especially if they demanded radical modifications to currently successful activities. The main reasons for inertia were power and political issues with the firms, complex bureaucratic processes such as time-consuming formal procedures, and the large number of management hierarchies that resulted in delayed decision making. For example, at a large automobile manufacturer, several different business units were working on biopolymer technology. However, these units were not allowed to share their results and experiences without prior formal consent of a number of their superiors. Since these formal approvals were very time-consuming, many researchers were reluctant to go through these bureaucratic processes. As a consequence, the decision to pursue biopolymer technology actively and to enter the market was unnecessarily delayed. In a firm active in the packaging sector, formalized responsibilities as well as bureaucratic and centralized procedures left no room for explorative spirit and innovative culture. As a result, many employees focused only on their regular work and did not "go beyond the usual."

Furthermore, the traditional polymer business, based largely on large-scale firms in relatively stable business environments, promotes administrative rather than innovative and explorative activities. Innovation is therefore often incremental with the aim of achieving cost reductions via process improvements and upgrades of existing products. For example, innovation activities at converters focus on reducing of cycle times in the injection molding process for semifinished or end products. At raw material suppliers, much of the research effort is aimed at modifying the material properties of existing petro-based polymers, for example, by improving heat stability or viscosity to achieve better processing performance.

The studied firms acknowledged the need for a different approach to develop and adopt biopolymer technology effectively. Biopolymer technology development needs to be supported by the main business activities to avoid opposition from individuals and departments involved in those activities. Furthermore, the biopolymer development activities need to create their own set of processes that tackle the uncertainty that surrounds biopolymers. Against this background, the study found four interesting examples of how firms in different sectors develop biopolymer technology.

Bootleg Research

In many studied companies, scientists were encouraged to conduct so-called bootleg research on biopolymers. Bootleg research is defined as research in which motivated employees, sometimes secretly, organize a part of the corporate innovation process (Augsdorfer 2005; Knight 1967). Normally, it is a bottom-up research activity, often not authorized by the management, but nonetheless with the intent to create benefits for the firm. Bootleg research is not in the companies' budget plans and has no official firm resources allocated to it. It can be found most often in preresearch, product and process improvements, troubleshooting, and purely scientific research (Augsdorfer 1996, 2005).

The results of the case studies showed two main mechanisms by which bootleg research helped firms to develop biopolymer technology successfully. In most firms, researchers who worked in the field of polymers had become aware of the new biopolymer technology through conferences, workshops, or scientific publications. However, research on biopolymers was often not in line with the traditional and officially planned technology trajectory of their firm. An official bootleg research policy helped scientists start informal research on this new technology without having to undergo approval processes to gain official legitimization. The researchers often conducted small-scale experiments in order to prove the feasibility of their approaches before submitting large-scale research proposals. This option gave bootleg researchers the possibility of challenging existing technology strategies within the firm without having to ask for corporate resources or being stopped by bureaucratic red tape.

Another mechanism by which bootleg research helped firms develop biopolymer technology involved anticipating and preparing to solve technological difficulties that would arise when the firm decided to develop biopolymers. These difficulties would arise because biopolymer technology requires a very specific set of resources and capabilities, for example, profound knowledge in biotechnology and process engineering, that the firms focused on petro-based polymers would not have. Yet, when bootleg researchers developed some of these new capabilities in their unofficial research, they were in a position to help solve technological difficulties the firms subsequently encountered. Such preparation also reduced time to market in some instances.

Research Alliances and Learning from Customers and Suppliers

Most of the companies studied were confronted by technical difficulties and market uncertainty as they developed their biopolymer technology. These difficulties were particularly severe for the smaller firms in the sample, making them less likely to succeed in developing, processing, and marketing this new technology independently. One strategy for addressing this problem was for these companies to form research alliances with other organizations.

Some firms collaborated closely with their customers and suppliers, mostly in the form of Research and Development (R&D) consortia, to increase their knowledge of future market requirements and possible technologies to fulfill these requirements. These firms also learned about biopolymers from other customer contacts. For example, when the management of one firm sought to increase revenues, they sent a team of engineers to meet with various potential customers. Their aim was to understand and solve the customers' problems by developing new petro-based products, but they also questioned them on their wider market needs. In this way, the firm got to know of their customers' interest in and need for products made out of biopolymers. Another firm worked very closely with suppliers that were considered innovative. These contracts helped management learn about new technologies that were previously unknown to the firm.

The establishment of official or de facto standard-setting or standard-promoting associations was important. Several of the studied firms formed alliances with other firms to reach agreement on common standards for biopolymers. These alliances were especially helpful in reducing technological uncertainty. The activities and commitments of other firms increased the confidence of the management that biopolymers were indeed a promising field to invest in. Agreements on common standards also helped in focusing on research activities and avoiding getting sidetracked. As a consequence, the common standards allowed firms to spend their R&D resources more efficiently.

A common characteristic of all reported alliances was their explorative nature. The firms were confronted by unfamiliar terrain and had to develop and adopt biopolymer technology that was often substantially different from either partner's current technology. To capitalize on the potentially distinct knowledge bases of these other firms, companies engaged in a learning process of trial, feedback, and evaluation. These alliances also enabled the firms to reduce costs and risks while enhancing technological predictability.

Cross-Functional Cooperation

The majority of the firms in the study relied on cross-functional cooperation for the development of biopolymer technology. Many studies have shown that improved cross-functional integration leads to improved development and adoption of radical innovations in terms of improved manufacturability and quality, reduced production costs, and fewer ramp-up problems (Karagozoglu and Brown 1993; Song, Neeley, and Zhao 1996).

All the firms studied used two main mechanisms for cross-functional cooperation in developing biopolymer technology. One mechanism is the accelerated exchange of information and its processing within the firms. In eight firms, the cross-functional team consisted of representatives from R&D, marketing, and production. They were able to tap a broad array of external information and new know-how. Involving marketing and production representatives in the team was particularly effective in facilitating the knowledge transfer necessary for the successful hand-off of the newly developed biopolymer innovation first to manufacturing for production and then to marketing and sales for distribution to the customers. These transfers were achieved by bringing employees with expertise about the proposed innovation together in face-to-face meetings. The close cooperation between the different functions involved led the team to a common understanding of the requirements and the content of what eventually evolved as the development focus. At converters, for example, the teams often also included employees from external suppliers such as raw material producers. These external supplier members acted as consulting team members rather than as core team members because they did not participate in the daily work. Generally, team members learned more about other disciplines and units in the firm and tended to develop new technical and other work-related skills faster because they worked across job functions. In the studied firms, teams were able to work on several biopolymer development issues in parallel.

Decentralization of decision making was a second mechanism by which the setup of cross-functional project teams helped in developing successful biopolymer technology. Many firms confirmed that in previous innovation projects, centralized decision procedures were one of the main problems in innovation processes. Decision making often took a long time during which projects got delayed and deadlines could not be met. This decentralization was achieved mainly through the use of lateral decision processes that were able to cut across the traditional vertical hierarchies of functional authority. In most firms, for instance, the project leader reported directly to the divisional head, which increased the chance of cooperation from all affected departments. Furthermore, the decision-making process was accelerated when team members did not have to report to various hierarchical levels. Successful teams, particularly those in large firms, benefited from having high levels of internal and external decision-making authority and from not being caught up by slow bureaucratic processes.

Most of the cross-functional project teams tended to be more effective and efficient in the biopolymer technology development process due to their decision-making authority. Team members were more committed to their work since they were able to negotiate with steering committees on all final decisions. In general, the cross-functional membership of the teams helped avoid the "not-invented-here" syndrome and the inertia that accompanies it.

Independent Project Houses

In recent years, large high-technology firms have increasingly established so-called project houses to keep up with smaller and more flexible competitors. A project house contains expertise from different areas and manages the innovation process, often from initiation through commercialization. The project house innovation process is characterized by extremely efficient usage of time by a small group of technical and managerial employees (Augsdorfer 2005). The goal of project houses is the development of new technologies, which can be achieved with minimal investments and which allows capacities to be adapted flexibly to meet evolving demands. A shorter time-to-market is often achieved through the systematic interaction with market research teams and potential customers, and by reducing managerial constraints and having a custom-designed workplace located near—but separate from—other firm facilities (Maine 2008). Top management often takes an active role in setting up project house operations and ensuring human as well as financial resources. It is necessary to isolate project housework from corporate bureaucracy. Demands for regular activity reports and delays in obtaining necessary funds and equipment—activities typical of traditional corporate life—would destroy their speed and flexibility (Augsdorfer 2005).

The results showed that raw material suppliers and automotive firms in particular were actively setting up project houses for the development of biopolymer technology. For example, at one large raw material supplier, the project house was seen as the major success factor in developing the company's biopolymer technology. Following the development of its first biopolymer ten years earlier, the company recognized that the various advantages of biopolymer technology, such as positive environmental impact, needed to be communicated to the company's customer base. Therefore, it built up a special organizational structure for the market introduction of this novel technology and established a project house as part of that structure. According to the responsible project manager, this new kind of organizational structure made the group much more efficient than their competitors. In this project house, most of the staff was recruited from units such as R&D, application technology, and manufacturing, and reported directly to the head of the polymer business unit. Marketing and sales staff were also recruited for this project, which was rather atypical for this firm. Traditionally, marketing and sales had been centrally located for all types of polymers and chemicals. The main mechanism used to overcome lock-in was giving project members total independence from the parent company in their decisions and budget issues. Thus, the head of the project house made decisions much faster than would normally be the case. All interviewees in this company confirmed that the project house was the most important success factor since it helped to "move things faster," a particular advantage in a large firm with over 100,000 employees.

For that company, the project house was the first of its kind. Other firms in the case studies were intentionally establishing project house operations to develop biopolymer technology because of good experiences in the past with this kind of innovation process, especially for other radical innovations. For example, at one large automobile company, project houses for the development and introduction of various different technologies had been very successful for a number of years. The main reason for their success was that project house employees were free from firm policies and politics. Employees of the new biopolymer project house in this company were mostly engineers and scientists with managerial experience, expertise, and technological know-how in polymer processing and biotechnology. They were selected from operating divisions and were on extended loan from them. The project house was located in rented quarters, physically separated from other facilities of the company. Although the project house was in walking distance from most of the corporate resources, such as various laboratories, visits by management were not frequent. In this case, the main mechanism against lock-in was the physical separation of the project house, which discouraged these visits and allowed the project house staff to concentrate fully on biopolymer development rather than on the next management review. Team-building among the project house staff was also facilitated by the physical separation from other units. The result was a unique integrated team that followed a clear vision in complete isolation from corporate inertia. The project house was designed and laid out to promote creativity and facilitate the innovation process for biopolymer technology. Project members were not tied to any bureaucratic red tape or influenced by firm politics. Strong support from the group executive and the general managers of the operating divisions was also recognized as a prerequisite for the success of this project.

Discussion and Conclusion

The replacing of petro-based polymers by biopolymers or other substitutes is an important step in moving toward a more sustainable world. The proliferation of biopolymer technology is a worldwide phenomenon that has increased rapidly since the early 1980s. In the past, biopolymers have usually been significantly more expensive and less efficient in their processing properties than their petro-based counterparts. However, increasing oil prices have made biopolymers more cost-competitive and have spurred research in molecular science and genetic engineering to improve the technology and technical performance of biopolymers.

Promising market developments indicate that biopolymers are about to leave their market niche. At a time when environmental laws are being strengthened worldwide, corporate environmental consciousness is increasing, public pressure for environmentally friendly technologies is high, and fossil fuel abundance is declining, a strong increase in biopolymer research activities is occurring and these factors are helping this new technology gain acceptance in a variety of industrial sectors.

However, biopolymers must continue to improve in price and processing properties. The need for improved properties is especially important for automotive applications. Since there are only a few suppliers with relatively small production volumes on the market, the overall availability of biopolymers must also be increased. Therefore, many market players are looking at raw material suppliers to provide the technology push necessary for the commercial breakthrough of biopolymers. But, since products made of biopolymers are still niche applications, large investments in research and development mean considerable financial risks for raw material suppliers as well as for converters.

However, applications made of renewable resources are no longer a mere lip service in the industry. Innovative firms see biopolymers as a chance to establish themselves as leading players in the new market. To gain the advantages of biopolymers, further research is required to optimize production by increasing the efficiencies of the different chemical and biotechnical processes involved. The key factors that will determine how rapidly biopolymer production and use expands in the coming years will be the depleting of oil reserves, the biodegradability of biopolymers, the use of legislative instruments supporting the use of biopolymers, the suitability of biopolymer material properties for converters, the versatility of applications, and the commercial viability of production and processing.

References

Augsdorfer, P. 1996. *Forbidden fruit: An analysis of bootlegging, uncertainty, and learning in corporate R&D*. Aldershot: Avebury.

Augsdorfer, P. 2005. Bootlegging and path dependency. *Research Policy* 34(1): 1–11.

Bastioli, C. 2005. *Handbook of biodegradable polymers*. Shawbury, UK: Rapra Technology.

Beucker, S., and F. Marscheider-Weidemann. 2007. Zukunftsmarkt Biokunststoffe. Berlin: Umweltbundesamt.

Busch, T., and V. H. Hoffmann. 2007. Emerging carbon constraints for corporate risk management. *Ecological Economic* 62(3–4): 518–528.

Crank, M., and M. Patel. 2005. *Techno-economic feasibility of large-scale production of bio-based polymers in Europe*. Sevilla, Spain: European Commission—Institute for prospective technological studies.

Dahlin, K. B., and D. M. Behrens. 2005. When is an invention really radical? Defining and measuring technological radicalness. *Research Policy* 34(5): 717–737.

Eisenhardt, K. M. 1989. Building theories from case study research. *Academy of Management Review* 14(4): 532–550.

Eisenhardt, K. M., and M. E. Graebner. 2007. Theory building from cases: Opportunities and challenges. *Academy of Management Journal* 50(1): 25–32.

Energy Information Administration. 2007. Forecasts and analyses. Washington, D.C.

European Bioplastics. 2006. Berlin.

International Energy Agency. 2007. *World energy outlook 2007*. Paris: OECD.

Kaplan, D. L. 1998. Biopolymers from renewable resources. Berlin; Heidelberg: Springer-Verlag.

Karagozoglu, N., and W. B. Brown. 1993. Time-based management of the new product development process. *Journal of Production Innovation Management* 10(3): 204–216.

Knight, K. E. 1967. A descriptive model of the intra-firm innovation process. *The Journal of Business* 40(4): 478–496.

Lee, S. Y., and S. J. Park. 2003. Economic aspects of biopolymer production. In *Biopolymers—general aspects and special applications,* ed. A. Steinbüchel. Weilheim: Wiley.

Maine, E. 2008. Radical innovation through internal corporate venturing: Degussa's commercialization of nanomaterials. *R&D Management* 38(4): 359–371.

Michael, D. 2004. Bioplastics supply chains—implications and opportunities for agriculture. Barton, Australia: Rural Industries Research and Development Corporation.

Miles, M. B., and A. M. Huberman. 1994. *Qualitative data analysis: An expanded source book* (2nd ed.). Thousand Oaks: Sage Publications.

Mohanty, A. K., M. Misra, and G. Hinrichsen. 2000. Biofibres, biodegradable polymers and composites: An overview. *Macromolecular Materials and Engineering* 267(1): 1–24.

Mohanty, A. K., M. Misra, and L. T. Drzal. 2002. Sustainable bio-composites from renewable resources: Opportunities and challenges in the green materials world. *Journal of Polymers and the Environment* 10(1/2): 19–26.

Paster, M., J. L. Pellegrino, and T. M. Carole. 2003. *Industrial bioproducts: Today and tomorrow.* Columbia: Energetics.

PlasticsEurope. 2009. *The compelling facts about plastics 2009: An analysis of European plastics production, demand and recovery for 2008.* Brussels, Belgium: PlasticsEurope.

Platt, K. D. 2006. Biodegradable polymers. Shawbury, UK: Rapra Technology.

Ren, X. 2003. Biodegradable plastics: A solution or a challenge. *Journal of Cleaner Production* 11(1): 27–40.

Runge, W. 2006. *Innovation, research and technology intelligence in the chemical industry.* Stuttgart: Fraunhofer IRB Verlag.

Sartorius, I. 2003. Biodegradable plastics in the social and political environment. In *Biopolymers*, ed. A. Steinbüchel: 453–472. Weinheim: Wiley.

Song, X. M., S. M. Neeley, and Y. Zhao. 1996. Managing R and D-marketing integration in the new product development process. *Industrial Marketing Management* 25(6): 545–554.

Vellema, S., R. van Tuil, and G. Eggink. 2003. Sustainability, agro-resources, and technology in the polymer industry. In *Biopolymers,* ed. A. Steinbüchel: 339–363. Weinheim: Wiley.

CHAPTER 12

Informal Waste-Pickers in Latin America: Sustainable and Equitable Solutions in the Dumps

Candace A. Martinez

This chapter discusses the present and future role of informal recycling in Latin America in contributing to countries' sustainability objectives as well as to firms' profitability and corporate social responsibility (CSR) goals. Considered one of the most undesirable of occupations, the task of sorting through and recycling refuse from city garbage dumps has undergone an important paradigm shift that has significantly affected the rules of the game for the "waste-pickers" or "scavengers" in many Latin American cities. Governments are increasingly aware that informal waste-pickers are providing a service that can and should be formalized. Private-public partnerships have developed as the civil sector (citizen groups and nongovernmental organizations or NGOs), the private sector (domestic and multinational enterprises), and the public sector (municipal and national governments) recognize the societal, environmental, and economic value of the waste-pickers' work. An examination of various aspects of the informal recycling sector in Latin America leads to the conclusion that stakeholders can work together to devise innovative approaches for participatory solid waste management that contribute to a more sustainable and equitable world.

> The idea is not to fight for solid waste but to have a win-win model that includes waste-pickers and the environment. The State should have incentives for the sustainability tripod, financial, social, and environmental. The psychological profile of the waste-pickers is of entrepreneurs, who we should be supporting.
>
> Cecilia Rodríguez, Former Environment Minister of Colombia (Rodríguez, 2008)

Sustainability will not be reached by technology alone, but by deep learning by individuals, groups, professional societies and other institutions.

H. Brown and P. Vergragt, 2008

Introduction

The task of sorting through and recycling refuse from city streets and open garbage dumps is undergoing an important paradigm shift. Scavengers (or informal waste-pickers/recyclers) around the world, especially those of Latin America, are forming cooperatives and associations to lobby for legitimacy, better working conditions, and the right to a fair wage for their work. Their voice is starting to be heard. For the first time ever, a small but well-organized group of waste-picker representatives attended the Conference on Climate Change in Copenhagen in December 2009 (COP15). As or more important, however, innovative partnerships are developing as the waste-pickers' demands resonate with the overlapping social, economic, and environmental objectives of the civil sector (citizen groups and NGOs), the private sector (domestic and multinational enterprises), and the public sector (municipal and national governments). Figure 12.1 represents the three overlapping sustainability objectives—social, economic, and environmental—of this "Triple Bottom Line" perspective.

Last year in Lima, Peru, for example, informal waste-pickers/recyclers formed the National Movement of Recyclers of Peru, supported in part by private enterprises. The organization currently has legislation in the Peruvian

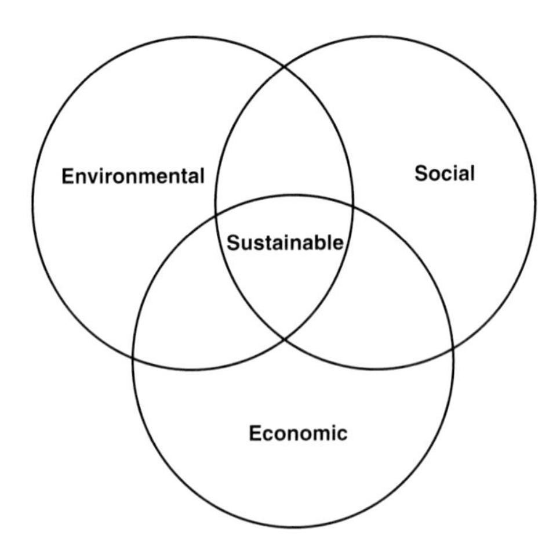

Figure 12.1 The Triple Bottom Line

Congress that is backed by the Environment Ministry because of its potential contribution to the country's sustainability goals of slowing down global warming (Chauvin 2009). Peru's informal waste-pickers have also taken an important first step toward legitimacy by referring to themselves as recyclers, not scavengers.

Similarly, Colombia's Constitutional Court ruled in April 2009 that the informal garbage pickers who live on or near dump sites were to be legally designated as self-employed entrepreneurs. An immediate consequence of the law is that the waste pickers' cooperatives can legally bid on waste-management concessions heretofore open only to private firm bids (*Economist* 2009). Cities and towns across Latin America are witnessing analogous grassroots movements as public and private actors reevaluate the important contributions the historically disenfranchised informal sector of waste-pickers is making to their sustainability needs.

The significant institutional changes occurring in the informal recycling sector in Latin America need to be understood by academics, policy makers, and practitioners alike. That governments will incorporate informal practices into formal laws has long been predicted by theory. The new institutional economic perspective, with its focus on real world conditions (Furubotn 2001), suggests that societal groups will create alternative institutions to provide the necessary economic governance when formal institutions in a society are weak and governments cannot provide effective property rights protection or enforce contracts (North 1990, 2005). These two forces exist in tandem, but the informal economic activities often supplant the formal actions over time, usually for the greater good (De Soto 1989; Rodrik 2003:10–16). All institutional change has winners and losers, however, and historical evidence suggests economic growth and prosperity have been more prevalent in those societies (e.g., Britain and the United States) that were responsive to informal sectors, promoted their survival, and were flexible enough to absorb their informal (de facto) practices into new (de jure) laws that better reflected reality.

Bringing waste-pickers, a formerly marginalized sector, within the legal fold augurs well for their economic future, giving them a sense of legitimacy, providing more opportunities for employment, and empowering them to demand improved standards of living. Moreover, governments are increasingly aware that waste-pickers are providing an economic service; shared sustainability goals can more readily be achieved when this informal sector of the working population receives formal status, and the resulting acknowledgment and support. As many experts in Latin America comment: It is a *gana-gana* (win-win) for everyone.

This chapter begins with a description of the formal/informal garbage collection and recycling sector in parts of Latin America. It then examines the institutional changes the sector has undergone—changes that potentially pave the way for investments that contribute to a more sustainable world. Next, it presents the benefits that associations of waste-pickers provide and reports on one of the most important associations of waste-pickers in Latin America, AVINA. After discussing the important role the national government of Brazil plays in the success of AVINA and other waste-picking associations, the chapter concludes with recommendations to the main stakeholders in the informal waste-collecting sector: government, private enterprise, and academia.

Informal Waste Collecting and Recycling in Latin America

Waste Collecting and Disposal: Two Models

To understand the world of informal recyclers (also known as waste-pickers or scavengers) in developing countries, and specifically in Latin America, a brief overview of the municipal solid waste collection and disposal system in developed countries is in order.

Figure 12.2 presents a conceptual schema of how the trash collection and disposal process operates in high/middle-income countries versus low-income countries. One of the most notable differences between the developing-country and developed-country models is that most towns and cities in the former dispose of their solid waste in open dump sites

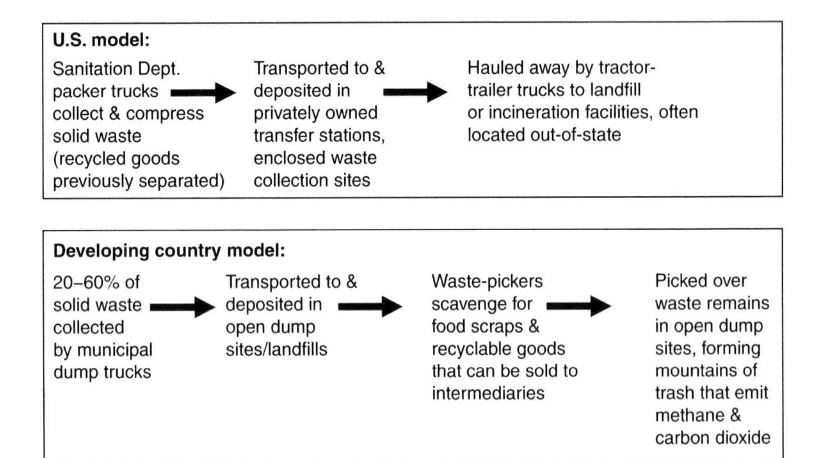

Figure 12.2 The value chain of municipal solid waste collection and disposal

Source: Devised by author from various sources

(although "sanitary" or enclosed landfills do exist in some locations), while in the latter countries open dump sites are used infrequently, mostly in small towns. Using the United States as a benchmark for how the industry works in advanced economies, private or Sanitation Department garbage trucks collect refuse, usually weekly, from residential and commercial locations at curbside, in alleys, or at other designated sites. Then they haul the solid waste to a transfer station, an enclosed warehouse-like facility, where it is dumped and stored until pick-up—normally within 24 hours—by large tractor-trailer trucks that transport the waste to a landfill where it is unloaded and covered. Landfills in the United States tend to be closed to the public, protected, even "secretive." This entire process, initial residential/commercial pick-up by sanitation department garbage trucks and eventual deposit in landfill sites (privately or publicly owned), can take anywhere from one to three days, depending on the distance from the central warehouse to the landfill. Royte (2005) provides a detailed account of the collection and disposal of solid waste in the United States. As Royte reports, some solid waste is incinerated in the United States, although this method is controversial because of its negative environmental spillover effects. Recycled goods (paper, aluminum, glass, et cetera) are separated upstream (by individual consumers, households, businesses), although the recycling process differs significantly from city to city and from neighborhood to neighborhood in the United States.

In contrast, the solid waste collection system in cities across developing countries follows a more primitive model. Although oversimplified, the fundamentals of the conceptual model in Figure 12.2 hold true for many urban areas in the developing world. Depending on the country and the city, anywhere from 20–100 percent of all solid waste is collected by garbage trucks and hauled to open dump sites (Medina 2007). These sites and their surrounding areas are home to thousands of informal waste-picker recyclers who scavenge, collect, and recycle glass, paper, aluminum, and various other goods. The informal waste-pickers gather around the trucks as they approach the site and vie for a more advantageous position from which to pick through the garbage as it streams out of the truck (Iwerks and Glad 2006). The scavengers collect whatever they deem of personal or "business" value—scraps of food or clothing for personal use or a wide variety of items for resale. The day's collection can be used/consumed immediately or saved for later resale to recycling intermediaries. Specialization is common. Some scavengers collect cardboard exclusively while others may collect metal only. At the end of the day, the trash that they reject contributes to the growing mountains of the city's daily refuse, adding to the public health and environmental risks that prevail in and around the open dump site.

The Lives of Informal Recyclers in Latin America

There are approximately 15 million scavengers in the developing world today, 3.8 million of whom can be found in Latin America (Medina 2009). These informal waste-pickers and recyclers are a group of people who collect, sort, and sell goods for recycling to small or midsize intermediaries. They often live on the street or in makeshift shanties, in or near the open dump sites, where they spend 10–12 hours daily, sorting through trash and recovering recyclable materials for resale. In some cities, dump sites are not open to the public and the waste-pickers use carts to pick up recyclables throughout the city, sleeping with them at night if they have nowhere to store them. Their abuse of alcohol and acts of violence tend to be higher than those of the general population, and they are usually perceived as unclean and a nuisance by authorities and middle-class citizens.

Collecting trash, whether in the street or in the garbage dumps, is a high-risk activity, both from a health and from an environmental point of view. Not only are transmittable diseases rampant, but fires erupt sporadically from the buried methane gas, dump truck accidents occur, and avalanches of trash sometimes claim victims. (See Cointreau 2006, for a detailed account of the panoply of diseases, injuries, and accidents that afflict workers exposed to solid waste sites in developing countries.) A 2005 fire in the open dump site in Guatemala City led to the government's enclosing the site with a wall, banning children from entry, and requiring that all adults entering the site carry government-issued ID cards that cost 45 Quetzales or about USD 6.00 (Iwerks and Glad 2006; informal conversations between author and experts in Guatemala City 2010).

Another risk concerns the waste-pickers' income streams. While an informal recycler can earn minimum wage or more, depending on the country and the market for items sold to recycling intermediaries, s/he is at the mercy of the global economy. The current financial downturn has negatively affected the price that middlemen pay informal recyclers for recycled commodity goods. According to the National Movement of Recyclable Materials Collectors in Brazil, a scavenger who was earning as much as $350/month for the resale of recycled materials saw his/her income drop by almost 50 percent to barely $200/month. A kilo of cardboard, for example, that sold for 22 cents on the dollar in September 2008 currently fetches about 6 cents. Even in good times, uncertainty is the hallmark of a waste-picker's income. While exact figures are difficult to obtain, a survey of 455 Brazilian "catadores" or waste-collectors (those who collect and sell trash) conducted in Pelotas, Brazil, in 2004 revealed that the average monthly income of an informal waste-picker was about USD 81.64, approximately half of what informal workers in another line of work earned and about equal to Brazil's minimum wage (daSilva et al. 2005).

> Waste-pickers' work is the perfect example of sustainable development.
> Martin Medina (Medina, 2008b:16)

Institutional Changes in the Informal Recycling Sector in Latin America

Waste-pickers live better and contribute more to the economy and to the environment when they are legalized and allowed to form informal recycling cooperatives or associations, as well as partnerships with public entities (Medina 2009). The evidence suggests that if the millions of scavengers in Latin America were to become legitimate on a broader scale, the overall waste collection system and the sorting of recyclables at the source (a cleaner, more efficient approach) would provide the tens of thousands of informal collectors in the larger cities a decent wage and a brighter future (Medina 2008a).

Several countries in Latin America have been pioneers in introducing institutional changes that foster this type of economic development as well as more ecological recycling and waste collection. The governments in Brazil, Colombia, Peru, and Argentina have replaced the de facto arrangement of this informal sector with a formal de jure arrangement, recognizing that the waste-pickers and other recycling workers in the informal economy are highly efficient environmental entrepreneurs to be supported, not ignored, or worse, ridiculed. In these countries, waste-pickers are now known officially as "entrepreneurs." The name change is not a trivial accomplishment. On one level, it enhances informal recyclers' self-esteem and paves the way for a more gradual public acceptance of their value and legitimacy in society. On a deeper level, the new nomenclature embodies a longstanding (and ongoing) grass roots struggle by informal waste-pickers to be taken seriously, to be allowed to practice their trade without fear of retribution or scorn, to be recognized as important and contributing members of society, and to be able to have a voice in future waste management partnerships with private enterprises.

Legal status, in principal, means that the waste-pickers can be self-employed, have social benefits, and as members of a legal organization can bid on government trash collection and recycling tenders, and/or be hired by private enterprises. The name change does not always work as anticipated, however. In Argentina, legally changing waste-pickers' title to entrepreneur did not produce the desired effect of addressing the problem of inefficient waste collection and recycling. The municipal government of Buenos Aires, Argentina, put into effect a law (Ley 992) in 2002 that legalized the *cartoneros* or garbage-dump scavengers. But in 2003, when the city's waste collection contracts expired, the same government excluded the newly legalized waste-pickers and their associations from being able to bid in the competition for new contracts. Only businesses were allowed to participate, despite the fact that it was estimated that the informal waste-pickers and recyclers were salvaging approximately 10 percent of the city's 5000 tons of daily solid waste (Schamber and Suarez 2007).

Associations in Latin America

Benefits of Informal Recycling Associations

Individual waste-pickers have little power vis-à-vis municipal governments and private enterprise. When they form collectives and associations, however, their unified voice gets noticed. These associations can influence four broad groups of constituencies: the waste-picker/recyclers themselves, the municipalities they help clean, the environment they help to make more sustainable, and the private enterprises that harness their recycling expertise and with whom they form partnerships.

First of all, the standard of living of the recyclers themselves can improve. Medina (2007) reports that scavenger recyclers in Nuevo Laredo, Mexico, can garner as much as three times that country's minimum wage. In Belo Horizonte, Brazil, after waste-pickers formed an association and partnered with the municipal government, their income increased from minimum wage to two-to-four times that amount (Dias 2000). Moreover, formal organization through cooperatives and associations augurs well for future opportunities. Organization leads to lobbying for legal, legislative, and procedural reform. The waste-pickers/recyclers can bid on municipal solid waste contracts, they can legally be subcontracted by private corporations, and they can exert more influence when meeting with government officials. The associations of waste-pickers can also learn from each other and leverage their knowledge, within countries and across countries, by diffusing their shared experiences, successes, and failures and gaining exposure through open dialogue at multiple forums: meetings, the Internet, and national and international conferences.

Second, materials recovery and recycling is the preferred option for all municipal waste management programs. Avoiding one ton of carbon dioxide (CO_2) emissions through recycling costs municipalities 30 percent less than doing it through energy-efficiency and 90 percent less than wind power, according to research by the waste-pickers' associations in India (Inclusive Cities 2009). Informal recyclers collect hundreds of thousands of tons of material from streets and city garbage dumps that get recycled and reprocessed into myriad products. The result is cleaner cities and more effective and efficient recycling. In Chile, it is estimated that 8.4 million cubic feet of waste does not go into municipal landfills and over 100,000 tons of CO_2 emissions are avoided, thanks to the efforts of informal recyclers there (Ingham 2009). In Curitiba, capital of the southern Brazilian state of Paraná, informal waste-pickers gather about 360 tons of garbage a day, representing approximately 70 percent of that city's paper, plastic, glass, and other recyclables (Inter Press Service 2003). And in Buenos Aires, more than 40,000 informal recyclers gather cardboard and other recyclable goods on the street with an economic value that could reach $178 million annually. It is calculated that informal recycling in Buenos Aires generates approximately USD 170 million a year in income for the recyclers (Medina 2007).

Third, the environmental advantages of waste-picker recycling are measurable and many. Informal recyclers (individually but more systematically when members of an association) save energy and trees through their resale of recyclable goods, to intermediaries who in turn sell in bulk (by the ton) back to industry for reprocessing. The Tellus Institute estimates that manufacturing processes that use recycled content produce 0.8 tons of CO_2 versus the 3.3 tons of CO_2 emitted when virgin materials are used (Tellus Institute 2008:12). Recycling aluminum cans specifically results in the greatest energy savings per ton since producing aluminum cans from virgin content is extremely energy-intensive. Using recycled aluminum in the process, however, requires relatively low levels of energy (EPA 2006: 98).

Recycling solid waste (inorganic materials) by scavengers reduces greenhouse gas emissions and produces other ecological benefits, as well. The decomposition process in open dumps and sanitary landfills generates methane, a greenhouse gas that is estimated to have 10–22 times more impact on climate change than does CO_2. Less waste reaching dumps and landfills translates into lower levels of LFG, a mixture of methane gas and CO_2. The hundreds of thousands of tons of material from streets and trash dumps that the informal recyclers collect represent waste that will not decompose in the sun, turn toxic, and reach the atmosphere. Relative to other methods of disposing of waste, recycling is more efficient and environmentally friendly than incineration. Even incinerating waste to produce electricity is not an attractive alternative, as waste incinerators emit 25 times more CO_2 per unit of electricity than do coal-fired power plants (de Nazareth 2009).

And, last, companies stand to gain as well from synergistic alliances with associations of waste-pickers even though the relationship between informal recyclers and the private sector has historically been contentious in the developing world where municipal governments have often displaced informal recyclers by subcontracting private companies to manage a city's solid waste management. Collecting, sorting, cleaning, and recycling materials reduces firms' demand for virgin materials, thus lowering their production costs, since manufacturing products using virgin raw materials requires double the energy usage relative to manufacturing products from recycled materials (Tellus Institute 2008). Environmentally conscious firms seek to incorporate recycling in as many stages of their value chain as possible. Wal-Mart, for example, uses approximately 1.2 billion plastic bags worldwide. By adopting a "recycled bags only" policy, the multinational has reduced by 30% its dependence on virgin resin, a derivative of petroleum. It has not only diminished the impact on the environment, but in Brazil alone, where Wal-Mart collaborates with the country's various waste-pickers' associations, it is estimated that the firm has saved USD900,000 in four months through its policy of recycled plastic bags (The ETHOS Institute of Socially Responsible Companies 2007: 22–23).

Natura Cosmética, S.A. of Brazil is another example. A direct-sales cosmetics company that entered the Colombian market in 2007, Natura Cosmética joined forces with the Recycling Association of Bogotá (ARB) to implement its in-house collection and recycling of all plastic bags its vendors use to deliver and pick up products at customers' homes. The firm employs approximately 75 ARB members full time and another 100 part-time (Natura Cosmética 2009).

AVINA and Its Initiatives in Latin America

One pan-Latin American organization, the AVINA Foundation, stands out as the overarching entity whose cohesive structure and organizational abilities are catapulting the hemisphere's informal waste-pickers and their causes into the political, social, and economic mainstream. This nongovernmental organization is at the forefront of forging strategic partnerships with NGOs, local and national governments, and private enterprises.

AVINA is an independent Latin American not-for-profit organization founded in 1994 by Swiss visionary and entrepreneur Stephan Schmidheiny, and continues to be funded, in large part, through a trust he created. Schmidheiny passionately believed that sustainable development was a necessary global goal that would be realized only through "alliances between successful and responsible businesses and philanthropic organizations that promote leadership and innovation," (AVINA 2010). AVINA has been instrumental in helping national cooperatives of waste-pickers across Latin America gain recognition by joining forces and lobbying their respective governments when perceived injustices take place (Fernandez 2009). In collaboration with Women in Informal Employment: Globalizing and Organizing (WIEGO), the Latin American and Caribbean Network of Waste Pickers, and the Latin American Waste Picker Network (LAWPN), AVINA played a key role in coordinating the Third Latin American Waste Pickers Conference/ First World Conference of Waste Pickers in Bogotá, Colombia, in March 2008, a milestone event that gathered, for the first time ever, waste-pickers and stakeholders from over 30 countries. The conference was underwritten by the Ford Foundation and Natura Cosmética, S.A.

Partnering today with almost one thousand leaders in civil society and in the business world, AVINA contributes to sustainable development in Latin America through its myriad programs and initiatives. Collaborations underscore this NGO's underlying principles as they apply to the informal recyclers of the region and demonstrate its focus on fostering socially responsible firms in Latin America. Building such firms is a necessary precondition for any socially inclusive, sustainability agenda, whether it addresses the needs of waste-pickers or those of other marginalized sectors of society.

Some of AVINA's partnerships include the following (AVINA Foundation 2009):

- The Bill and Melinda Gates Foundation is working with AVINA to create thousands of jobs and improve conditions for workers and the families of waste-pickers in six countries of South America. The Foundation has pledged $3 million for a four-year project to lend general operating support to Latin American recycling cooperatives (The Bill and Melinda Gates Foundation 2010). AVINA reports that leaders from the informal recycler movement were directly involved in the design of this project.
- AVINA developed a joint project named Program *Cata-Ação*/ "Socioeconomic Integration of Collectors of Recyclable Materials" in 2008 with the Inter-American Development Bank, Brazil's Ministry of Social Development, the Brazilian National Movement of Waste Pickers, and several companies. The project, which is funded with US$7.9 million across five communities in Brazil, contributes directly to the economic health of informal waste-pickers/recyclers and their

families. The anticipated outcome is that the project, if successful, will be imitated in other cities in Brazil and across Latin America.

- In Argentina, AVINA has partnered with the business sector by creating the National Movement of Businesspersons for Social Responsibility, a group that works with the Argentinean Institute for Corporate Social Responsibility to promote business strategies for reducing poverty and increasing corporate social responsibility (CSR) throughout Latin America. Bolivia, Colombia, Nicaragua, Paraguay, and Peru have established similar CSR initiatives through their national CSR associations: the Bolivian Corporation for CSR, the Colombian Council for Corporate Responsibility, the Nicaraguan Union for Corporate Social Responsibility, the Paraguayan Association of Christian Businesspersons, and Peru 2021. Approximately 900 Latin American companies, employing almost 200,000 people, belong to these associations.

Government Role in Brazil

From Brazilian President Lula downward, the various levels of government have openly embraced, and more important, financed the informal waste-pickers of the country. During a fall 2009 waste-collector event held in Sao Paolo, President Lula pledged to the 1500 waste-pickers/informal recyclers in attendance that a US$128 million line of credit for recycling cooperatives and associations would be created. Actions like this one are creating an environment in which private corporations' access to legally organized associations of recyclers and their services has been encouraged to flourish (The AVINA Foundation, 2009). The Brazilian Business Commitment for Recycling (CEMPRE) is one such organization. CEMPRE was founded in 1992, and is supported by private companies such as Natura Cosmética, Nestlé, Unilever, Alcoa, Wal-Mart, Kraft, and Coca-Cola. Its stated objectives are to raise public awareness about the importance of reducing, reusing, and recycling waste (the 3Rs) through promoting integrated management of municipal solid waste and recycling after use (CEMPRE 2010). The group supports the work of the informal waste-pickers' associations around the country and has found interesting and inventive formulas in which to partner with them for solid waste management projects. The following are representative examples of CEMPRE firms and waste-pickers' association collaborations:

- In the state of Bahia, Wal-Mart has entered into a partnership with the waste-pickers of that state to install and run recycling-at-source facilities for its customers at different waste-collection sites. Wal-Mart has also subcontracted with associations of waste-pickers to handle waste collection, disposal, and recycling at Wal-Mart's annual industry events and conferences.

- During the 2006 Carnival celebrations in Rio de Janeiro, Coca-Cola hired sixty waste-pickers from eight recycling cooperatives to collect trash and sort recyclable material in the streets where the samba parades were held. The waste-pickers sorted the recyclable materials in different garbage transfer points in the city, with the final tally of recycled material reaching 6.8 tons. (http://www.thecoca-colacompany.com/ourcompany/wn20060428_recycling.html).
- Since its founding in 1969, Natura Cosmética, S.A. has been a pioneer in fostering environmental awareness. Recognized by the U.N. for its contribution to supporting the environment, Natura uses native-grown materials from Brazil's vast biodiverse regions and was the first direct-sale firm to produce and sell refills of its cosmetics. A recycling system it established in its Recife, Brazil, operations provides an illustrative example of an industrial ecosystem, one that uses the materials from one industrial process as raw materials in another industrial process. In its Recife plant, Natura's direct sellers deliver all customers' orders in recycled plastic bags. The bags are then given back to the sellers when the customer places a new order, picked up by waste-pickers, and delivered to the association's transfer station where they are sorted, cleaned and sold to an industrial recycler. Natura replenishes its stock of plastic bags for new orders from the industrial recycler whose supply of newly recycled bags closes the positive feedback-loop process.

The move toward greater corporate accountability and social responsibility over the past decade has matured into innovative capitalism that uses market tools to solve social problems, and inclusive business approaches that generate wealth for the underserved majority.

Brizio Biondi-Morra (Chairman, The AVINA Foundation, 2008)

Conclusion

The significant institutional changes occurring in the informal recycling sector in Latin America have implications that are inextricably linked to broader sustainability concerns. The rules of the game are changing. Some governments in Latin America are awarding formal recognition and legal status to waste-pickers. The new recycling entrepreneurs have formed collectives and associations, their contributions to a greener environment are starting to become recognized, and they are gradually gaining legitimacy in society. Informal institutions, belief systems, and values change at a slower rate than do formal institutions, however (North 1990). The legal name change from waste-picker to recycler or entrepreneur represents a formal institution; it can happen overnight. A society's attitude toward and acceptance of its disenfranchised groups will take longer to change.

Although their full incorporation into the social fabric of a country may never be attained, and the level of acceptance of informal waste-pickers and recyclers as contributing members of society remains to be seen, this chapter has argued that they, at least, represent a significant contribution to green waste management and recycling practices. In addition, they form an integral part of the solution to responsible waste management in the developing world. As we go forward, the challenge is to save the planet's capacity to support our own and other species in an environmentally, socially, and economically responsible way. The evidence suggests that the current trend in legitimizing the informal waste-pickers in Latin America will continue. Innovative public/private/civil society partnerships with this historically disenfranchised population may prove to be a sustainable solution for the planet, a profitable solution for domestic and multinational firms, and a more equitable solution for the bottom-of-the-pyramid scavenger-recyclers. If we accept the premise that waste reduction and recycling offer lower environmental impacts than landfilling or incineration and consume less net energy than other solid waste management techniques, actionable recommendations can follow.

Recommendations

Public Policy. Governments everywhere need to recognize the crucial role that recycling plays in achieving environmental goals and public policy in developing countries need to proactively support the work of waste-pickers. Research has indicated that to unlock the potential of informal recyclers as valid contributors to a sustainable world, they must be incorporated into a society's legal framework (Medina 2007). National policy and legislation should explicitly recognize informal recyclers as key stakeholders in sustainable waste management systems. Municipalities should develop strategies to involve informal recyclers and their collectives and associations into waste management policy-making processes. Partnerships with other governmental groups, NGOs, and waste-pickers' associations form the foundation for an intelligent, enduring, inclusive sustainability policy that can contribute to meeting global warming challenges for this generation and future generations.

Governments at all levels also need to rethink the short- and long-term sustainability contribution of waste-to-energy projects. Not only in Latin America but elsewhere in the developing world, the environmental value and the social impact of these operations have been called into question. As the Institute for Policy Studies reported, the livelihoods of India's waste-pickers could be endangered if the government or private sector continues to build waste-to-energy plants, facilities that, if meeting the criteria of the Kyoto Protocol's Clean Development Mechanism plan, can earn lucrative carbon credits for trading or selling in the open market. Moreover, it is debatable whether the solid waste in countries like India makes incineration sense. According to Neil Tangri of the Global Alliance for Incinerator Alternatives, Delhi's garbage contains very little burnable matter because most of the paper and plastic are removed and recycled by the city's waste-pickers (Wysham 2008).

Private Enterprise. Many businesses are becoming leaner and greener recycling machines as they answer the call to arms for sustainability. If corporations are truly serious about CSR, as most claim to be, then private enterprises should take advantage of the efficiency and qualifications of waste-pickers' associations in the developing world by integrating them into a participatory solid waste management process. Firms can readily tap into the expertise of waste-pickers' associations in countries where these associations have a long history of crusading for and winning legitimacy, as they have in Brazil, Colombia, and other Latin American countries. Members are well organized and efficient, and the services they offer can be customized to meet the particular demands of a company. The holistic Natura Cosmética model shows that when corporate commitment is genuine, synergistic collaborations with waste-pickers' associations can reach new heights. Resources can be allocated to mutually beneficial, innovative solutions that incorporate informal recyclers into different stages of a firm's value chain, wherever it makes most economic sense. While the positive feedback loop approach that Natura Cosmética has been able to implement may be impractical for some companies to duplicate, the possibilities for other collaborative approaches are abundant. In addition to seeking out partnerships with waste-pickers' associations, private industry should team up with NGOs and government groups who are knowledgeable about the sector and can offer guidance. Domestic and multinational companies also need to support business associations, like CEMPRE in Brazil, that promote alliances with waste-picker and other marginalized groups in society. Joining a business association of this type reflects a firm's serious commitment to sustainability throughout the value chain, from the construction and operation of its facilities to the design and manufacturing of its products, to the use and disposal of its products, and to its active involvement in the community.

Academia. Experts forecast that worldwide waste will double by 2030, and that proportionately more waste will be generated by developing economies (*Economist* 2009). If a necessary condition of good scholarship is to ask pertinent questions that "address important issues and/or challenge existing beliefs" (Ordoñez, Schweitzer, Galinsky, and Bazerman 2009: 84), then cross-disciplinary research is a prerequisite if we are to ask the tough questions and illuminate possible ways forward. Social scientists—management scholars, political scientists, environmentalists, sociologists, economists, historians, geographers—all have a role to play in examining waste collection and recycling phenomena with their specialized lenses. Research can provide incremental contributions that add to our general knowledge base, extending extant theory to explain phenomena and predict possible outcomes. There is a dearth of studies on waste-pickers, informal recycling, and the recycling industry in developing countries. Academics need to collect and analyze data systematically using multiple methodologies (quantitative, qualitative, area studies, case studies, triangulating methodologies) and to apply theoretical insights to questions such as: Who are the waste-pickers? What are the different kinds of recycling work currently being conducted in the developing regions of the world? What are the direct and indirect effects waste-picker recyclers have on society and on greenhouse gas emissions? What percentage of informal recyclers join associations, and what are their reasons for belonging or not belonging? And what are the differences in the answers to these questions across countries? We need further studies to quantify the amount of material diverted from landfills by recyclers and to calculate the contribution that these diversions make toward sustainability and the life of the landfills.

From a corporation's perspective, we need to know whether those companies that partner with waste-pickers actually perform better over the long-term and whether their models provide sustainable competitive advantage. If so, how is that advantage measured and what incentives can encourage firms (and governments) to recognize and integrate, not ignore and exclude, the value waste-pickers and their associations can bring to an integrative solid waste management process and to the reduction of greenhouse gases? It is hoped that when more research is conducted by impartial scholars, the findings will be used by public and private stakeholders in sustainable waste management systems of the developing world to design and implement more inclusive and transformative approaches.

References

AVINA Foundation. 2009. Waste pickers summon the president of Brazil, mayors and businessmen. http://www.avina.net/web/siteavina.nsf/0/15C2BD6915487CB90325765F0043F8E0? OpenDocument&idioma=eng&sistema=1&plantilla=3&flag=Home& (accessed February 23, 2010).

———. 2010. http://www.informeavina2008.org/english/whoweare.shtml (accessed February 23, 2010).

Bill and Melinda Gates Foundation. 2010. http://www.gatesfoundation.org/Grants-2009/Pages/AVINA-Americas-OPP1007399.aspx (accessed January 23, 2010).

Biondi-Morra, B. 2008. The AVINA Foundation. http://www.informeavina2008.org/english/president_message.shtml (accessed February 23, 2010).

Brown, H., and P. Vergragt. 2008. Bounded socio-technical experiments as agents of systemic change: The case of a zero-energy residential building. *Technological Forecasting and Social Change* 75: 107–130.

CEMPRE. 2010. http://www.cempre.org (accessed February 21, 2010).

Chauvin, L. 2009. Peru's scavengers turn professional. *Time Magazine,* 10 February.

Cointreau, S. 2006. Occupational and environmental health issues of solid waste management; special emphasis on middle- and lower-income countries. The International Bank for Reconstruction and Development/The World Bank. http://www-wds.worldbank.org/external/default/WDSContentServer/WDSP/IB/2007/07/03/000020953_20070703143901/Rendered/PDF/337790REVISED0up1201PUBLIC1.pdf. Accessed 2 September 2009 (accessed December 5, 2009).

DaSilva, M., A. Fassa, C. Siqueira, and D. Kriebel. 2005. World at work: Brazilian ragpickers. *Occupational and Environmental Medicine* 62: 736–740.

de Nazareth, M. 2009. Waste pickers: Silent friends of the polluted earth. *Deccan Herald,* 3 September. http://www.deccanherald.com/content/d17818/waste-piockers-silent-friends-polluted.html (accessed September 22, 2009).

De Soto, H. 1989. *The other path*. New York: Basic books.

Dias, S. M. 2000. Integrating waste pickers for sustainable recycling. In *Planning for Integrated Solid Waste Management—Collaborative Working Group Workshop, 18–21 September, 2000*. Manila, the Philippines. http://www.docstoc.com/docs/DownloadDoc.aspx?doc_id = 20395589 (accessed February 2, 2010).

Economist. 2009. Muck and brass. U.S. Edition. 28 February. http://www.lexisnexis.com/us/lnacademic/results/docview/docview.do?docLinkInd=true&risb=21_T7199681467&format=GNBFI&sort = BOOLEAN&startDocNo=1&resultsUrlKey=29_T7199681470&cisb 22_T7199681469&treeMax = true&treeWidth=0&csi=7955&docNo=3 (accessed July 14, 2009).

EPA (The U.S. Environmental Protection Agency). 2006. Solid waste management and greenhouse gases: A life-cycle assessment of emissions and sinks. http://www.epa.gov/climatechange/wycd/waste/downloads/fullreport.pdf (accessed March 1, 2010).

The ETHOS Institute of Socially Responsible Companies. 2007. Vinculos de negocios sustentaveis em residuos solidos. http://www.ethos.org.br/_Uniethos/documents/VincSust_res_sold_A4.pdf (accessed January 14, 2010).

Fernandez, L. 2009. Latin American waste picker network. In *Refusing to be cast aside: Waste pickers organising around the world*, ed., M. Samson. Cambridge, MA: Women in Informal Employment: Globalizing and Organizing (WIEGO), pp. 44–48. http://www.inclusivecities.org/pdfs/Refusing%20to%20be%20Cast%20Aside-Wastepickers-Wiego%20publication-Chap3.pdf (accessed February 10, 2010).

Furubotn, E. 2001. The new institutional economics and the theory of the firm. *Journal of Economic Behavior and Organization* 45: 133–153.

Inclusive Cities. 2009. Wastepickers and climate change. http://no-burn.org/downloads/wastepickers-CC-EN.pdf (accessed February 2, 2010).

Ingham, R. 2009. Wastepickers of the world unite at climate talks. *Agence France Presse*. http://www.france24.com/en/node/4943441 (accessed February 15, 2010).

Inter Press Service. 2003. Eco-briefs—Brazil: Recyclers demand wages. http://www.tierramerica.info/nota.php?lang=eng&idnews=1561&olt=218 (accessed January 12, 2010).

Iwerks, L. (Producer), and M. Glad (Director). 2006. *Recycled Life*. (Motion Picture) Canada: Leslie Iwerks Productions.

Medina, M. 2007. *The world's scavengers: Salvaging for sustainable consumption and production*. Lanham, MD: AltaMira Press.

———. 2008a. The informal recycling sector in developing countries. *Gridlines*, Note No. 44: 1–4.

———. 2008b. Waste pickers without frontiers. In *Conference proceedings from the first international and third Latin American conference of waste-pickers*. Page 16. http://www.wiego.org/reports/WastePickers-2008.pdf (accessed September 12, 2009).

———. 2009. Global recycling supply chains and waste picking in developing countries. United Nations University: World Institute for Development Economics Research.

http://www.wider.unu.edu/publications/newsletter/articles/en_GB/12-2009-wider-angle-1/) (accessed January 28, 2010).

Natura Cosmética. 2009. Report by Natura Cosmética. Available from the author upon request.

North, D. 1990. *Institutions, institutional change and economic performance*. New York: Cambridge University Press.

———. 2005. *Understanding the process of economic change*. Princeton, NJ: Princeton University Press.

Ordoñez, L., Schweitzer, M., Galinsky, A., and Bazerman, M. 2009. On good scholarship, goal setting, and scholars gone wild. *Academy of Management Perspectives* 23 (3): 82—87.

Rodriguez, C. 2008. Waste pickers without frontiers. In *Conference proceedings from the first international and third Latin American conference of waste-pickers*. Page 17. http://www.wiego.org/reports/WastePickers-2008.pdf (accessed September 12, 2009).

Rodrik, D., ed. 2003. *In search of prosperity: Analytical narratives on economic growth*. Princeton, NJ: Princeton University Press.

Royte, E. 2005. *Garbage land: On the secret trail of trash*. New York: Little, Brown & Company.

Schamber, P., and F. Suarez. 2007. *Recicloscopio*. Buenos Aires: Prometeo Libros.

Tellus Institute. 2008. Assessment of materials management options for the Massachusetts solid waste master plan review: Final report appendices. http://www.mass.gov/dep/recycle/priorities/tellusmma.pdf (accessed February 15, 2010).

Wysham, D. 2008. The waste-pickers of Delhi. Institute for Policy Studies. http://us.oneworld.net/article/kyoto-protocol-endangers-indias-waste-pickers (accessed January 20, 2010).

CHAPTER 13

Sustainability of the Wine Industry: New Zealand, Australia, and the United States

Anatoly Zhuplev and Cynthia Aguirre

Introduction

Wine making is a mature and sophisticated industry with France, Italy, and Spain being the world's top producers, closely followed by the United States and Australia. Most of these major wine-producing countries also tend to be the largest and most mature wine markets with limited consumption growth potential. The increasing demand for wine from growing markets such as Russia, China, India, and the United States is expected to drive future growth in the industry (Interview with Philip Gregan, New Zealand Winegrowers, May 22, 2009).

Grape-growers and winemakers have a long history of caring for their land and preserving it for future generations as an inherent part of the industry "DNA." However, due to climate change, the urgency to minimize environmental impacts in the industry is especially acute. Along with environmental concerns, sustainability is also important in keeping winemaking competitive and thus sustainable in a business sense. Although consumer demand for sustainable wines has proved inconclusive thus far, it is a step in the right direction.

This chapter presents the findings of a comparative study of sustainability in the wine industry in New Zealand, Australia, and California—California is responsible for 90 percent of the wine production in America and is used as a representation of the U.S. wine industry. Sustainability in the wine industry can be assessed by observing general trends and subsequently investigating the best practices, from grape-growing techniques to packaging, that are being adopted by the winemakers and other wine-industry-related stakeholders within each country.

Effects of Climate Change

Both the average growing-season climate and temperature extremes affect vine and fruit development (Cahill and Field 2008). With the rise in temperature of between 1°C and 3.5°C anticipated in the next hundred years, the distinct styles and characteristics attributed to a specific wine region will change. With a rise in temperature, a correlative rise in humidity will also ensue as the moisture-carrying capacity of the air increases. As a result, some thin-skinned varieties of wine grapes will suffer due to mildew and moulds. Warmer temperatures also trigger heavier, short periods of rainfall. This situation typically results in less water absorbed into the ground (especially in sloping vineyards) and a decrease in water availability for the vineyard (Jones et al. 2005).

Warmer temperatures will mean lower acidity, which gives a wine its "bite" and lighter taste and feel, as well as a higher sugar content in the grapes. Through fermentation, sugar converts to alcohol, which in turn gives wine its heaviness. Therefore, as temperatures rise, wines will have fuller and heavier taste. While some people prefer rich and heavy wines, many of the classic styles rely on their acidity. Climate change not only poses a threat to a wine region's distinctiveness, but also to its viability, marketability, and even its mere existence. The development of new winegrowing regions in cooler climate zones and at higher latitudes is a highly probable response to inevitable climate change. Already, English wines are growing in strength as the "improving" climate makes winemaking in England more viable. Thus the wine industry is likely to feel the influence of climate change well before other agricultural industries (Jones et al. 2005).

Greenhouse Gas Protocols and Carbon Footprint Measurements

A carbon footprint is "the total set of greenhouse gas emissions caused directly and indirectly by an individual, organization, event, or product" (Carbon Trust 2010). More clearly, it is the measure of the impact of human activities on the environment with respect to climate change (Carbon Footprint 2010). The Wine Institute of California, New Zealand Winegrowers (NZWG), South Africa's Integrated Production of Wine program, and the Winemakers Federation of Australia have joined forces to create the International Wine Industry Greenhouse Gas Protocol (Global Wine Industry 2008). Its goal is to provide the wine industry with a free greenhouse gas protocol and calculator to measure the carbon footprints of wineries and vineyards of all sizes (Wine Institute 2008).

This methodology looks at three types of emissions in the winegrowing and production process: (1) direct emissions from the fuel used in water heaters, frost-fighting equipment, boilers, tractors, trucks, and harvesters, (2) indirect emissions from the production of purchased electricity, heat, or steam, and (3) indirect emissions from all products/activities that are purchased from/connected with other companies (e.g., fertilizers, packaging materials, and the transportation of purchased products to the winery and/or the transportation of wine products to the point of sale) (Wine Institute 2008).

The International Carbon Footprint Calculator is an innovative tool facilitating sustainability by giving wineries and vineyards the ability to calculate their own carbon footprint (New Zealand Winegrowers 2008). The downloadable online calculator is available for free and was developed through a partnership between wine industry organizations in New Zealand, Australia, South Africa, and California. While there are no binding regulatory requirements regarding levels of carbon emissions at this point, there was a strong demand for an internationally credible system (New Zealand Wine 2008).

Environmentally Conscious Approaches to Wine Making

Organic

Organic farming emphasizes reliance on natural fertilizers and pest control systems instead of synthetic materials (fungicides and pesticides). Under the organic wine concept, the winery also uses natural yeasts, minimal filtration, and fining materials. There are currently no mutually recognized international standards defining organic wine. Therefore, producers who grow certified organic grapes but process them conventionally by adding yeasts, bentonite, egg whites, gasses (N_2, CO_2), or other additives—or where the processing is not certified—generally use terms such as "wine made from organically grown grapes" on bottle labels (Organic Wine 2009).

Australia has well-developed standards for organic winemaking, and their "organic" wine is internationally recognized and welcomed in the industry (Organic Wine 2009). Three organizations certify almost all of the organic wine and grape production in Australia: the National Association for Sustainable Agriculture, Australia Ltd. (NASAA); Australian Certified

Organic (ACO); and the Biodynamic Research Institute (Organic Wine 2009). Processing standards for organic wine in Australia do permit some additives and processing aids, in contrast to New Zealand, and noticeably the United States where there are no legal organic processing standards (Organic Wine 2009). Despite some disagreement about whether wines produced organically yield the same taste and quality, wines processed under the strictest standards allow winemakers to market a "vegan" wine (Organic Wine 2009).

Biodynamic

Similar to organic farming, biodynamic farming is free of synthetic pesticides and fertilizers. However, biodynamic farming goes a step further and encompasses the whole winemaking process to ensure its sustainability in the long term. The fundamental principle of biodynamic farming is that the farm is managed as a living organism and as a system with a high degree of self-sufficiency in all aspects of biological habitat. Fertility and feed arise out of the recycling of the organic material the system generates; avoidance of pest species is based on its intrinsic biological vigor and genetic diversity; water is efficiently cycled through the system (About Biodynamic Agriculture 2009; Barker, Lewis, and Moran 2001).

A biodynamic farm should be regenerative rather than degenerative, and its important environmental value is that it does not depend on the mining of the earth's natural resource base. The biodynamic method of winemaking strives to align all of the factors inherent in a living farm system in a harmonious manner, thus leaving behind the lightest carbon footprint.

Sustainable

Sustainable winegrowing leaves the farm in at least as good, or perhaps better, shape for the next generation. It is development that meets the needs of the present without compromising the ability of future generations to meet their own needs (Ohmart 2009). Sustainability incorporates three main goals: environmental stewardship, economic profitability, and social and economic equity. Some of the benefits include long-term viability of land and business; improvement in wine quality; and enhanced relations with communities, consumers, and regulators (Potential Benefits 2009).

Sustainable winegrowing encompasses the soil, water, grapes, air, energy use, areas not farmed, family, employees, community, and economics. It is best viewed as a continuum where there is always room for improvement (Ohmart 2009). Sustainable farmers are doing their best to give back to the community and environment, while at the same time pursuing their business priorities. Sustainable farming may occasionally use synthetic materials, but only the least harmful and only when absolutely necessary. The goal is healthy and productive soil that produces healthy vines and will continue to do so for future generations (Osborn 2009). The latest statewide Certified California Sustainable Winegrowing (CCSW) program certification standards demonstrate that being certified as "sustainable" is not a cover-up allowing winemakers to use chemicals occasionally while still claiming to be "green." Truly sustainable winemaking may in some cases surpass the requirements of organic certification. Where organic farming talks about pesticide use, sustainable farming goes farther and tackles issues that are otherwise unaddressed with organic certification—such as wind power, water conservation, employee benefits, materials handling, and providing natural habitat for birds and predators (Nigro 2010).

Best Sustainable Practices in Vineyards

Canopy Management

The upper leaf-and-fruit-bearing portion of the grapevine is called the canopy. Ensuring that the grapes receive the proper amount of sunlight is essential for producing quality wines and simply keeping the process sustainable. One way to manage the temperature change is through canopy management and vineyard layout. More foliage may shelter the vine. There are a number of methods for creating cooler canopies (Jones et al. 2005). Ongoing advances in trellising the vines, positioning the shoots, and manipulating the canopy have resulted in greater intensity of fruit flavor in the finished wines (Modern Viticulture 2002). This in turn benefits the wine quality and reduces the need for chemicals, as well as electrical and mechanical equipment, making the winemaking process more sustainable and cost-efficient.

Technology and Mechanization

Recent advances in technology and mechanization have changed the way wine grapes are farmed. Satellite imagery that detects vine deficiencies, mechanical harvesters leaving under-ripe berries on the vine, weather stations in vineyards, and electrostatic sprayers limiting the amount of pesticide applied are all technologies that give growers powerful tools and information for producing world-class wines and contributing to sustainability of the grape-growing process (Modern Viticulture 2002).

Pest Management

Some vineyards rely on traditional strategies in their pest management, such as using sheep in vineyards as a green alternative to tractors. Sheep trained for an aversion to grape leaves aid with weed control, and thus have an increasingly important role in the vineyard ecosystem in deterring pests and keeping vines healthy (Trained Sheep 2007). Instead of utilizing machines or herbicides, sheep can do the job of several workers in less time, with less environmental damage, and often with better cost-efficiency compared to physical manpower.

Fostering Collaboration and Education on Sustainability

Coopetition—a Cooperative Strategy

"Coopetition" means a combination of (1) competition and (2) cooperation. As sustainability continues to gain momentum, and wine companies make this a part of their "infrastructural" scope, it can be best achieved through collaborative efforts on a local, national, and international level. Coopetiton is especially useful when firms competing within the same industry are faced with similar challenges; thus, working together can help ensure goal accomplishment in both sustainability and their respective bottom lines.

The NZWG is an example of collaboration at its best. The association provides extensive research and valuable data on various topics, ranging from wine production and sustainability practices to export and import of historical information. A subsection of NZWG, the Sustainable Winegrowing New Zealand (SWNZ) is a program that has been in place for over ten years, providing resources and leading initiatives in meeting the demand for sustainably made wines. SWNZ has earned an international reputation for high-quality and environmentally responsible wine production. Without their efforts in sharing information, providing advice and support to all firms big and small, few winemakers would have the resources or capital to gather such research on their own (Interview with Philip Gregan, New Zealand Winegrowers, May 22, 2009).

The California wine industry, the world's fourth-largest wine market with nearly 3,000 wineries (Cone 2010), has also been striving for collaborative information sharing on practical steps toward sustainability. The California Association of Winegrape Growers (CAWG), for example, represents over 50 percent of the grape tonnage used in producing wine in California. This statewide association serves as an advocate for farmers in providing education, research, and leadership on policies surrounding sustainable farming (CAWG 2007).

The new statewide CCSW program introduced in 2010 helps define sustainability, giving it equal weight with organic and biodynamic certification. The CCSW certification grew out of the 2002 Sustainable Winegrowing Program launched by the Wine Institute and the CAWG to educate their members about best environmental practices and to encourage them to assess the need for improvements within their own businesses.

The assessments account for 46 percent of California's 526,000 vineyard acres and 59 percent of the state's 240 million-case production. Rather than requiring all participants to adhere to the exact same practices, the CCSW program allows them to concentrate on the goals that best address the specific environmental and community priorities in their region. Wineries and vineyards start by evaluating their operations against the 227 best practices in the Code of Sustainable Winegrowing Practices Self-Assessment Workbook. Eligibility for certification depends on meeting 58 prerequisites including criteria related to air and water quality, water conservation, energy efficiency, reduced pesticide use, and the preservation of ecosystems and animal habitats. An approved third-party auditor paid by the company must then verify the accuracy of the scores and the practices (Nigro 2010).

Collaboration in sustainability can also be driven by culture. The New Zealand winemakers, for example, have characterized their collaboration as part of their culture and identity, maintaining friendly and helpful relationships between competitors. Even though each company wants to be successful in its own way, no one wants to spoil the positive image the New Zealand wine industry has overseas (Gabzdylova, Raffensperger, and Castka 2009).

Conversely, Australia's wine industry, much larger in scope than New Zealand's, requires continued leadership from research agencies, government, and winemaker associations in order to establish a collaborative spirit among competitors (Aylward and Clements 2008). They need more centralized "hubs" of information sharing, decision making, and resources available that will aid in creating more coopetition for marketing sustainability practices in Australia. The system also needs to be more inclusive for smaller wineries, especially since they are being the most innovative in maximizing their limited resources (Aylward and Clements 2008).

Consumer Education

While most current programs on sustainable wines focus on educating winemakers, very few focus on the consumer. Some experts believe that most consumers do not typically associate wine growing as being a polluting business; therefore, choosing environmentally friendly wines is not "on the front of their mind" (Reign of Terroir 2009). In general, a large percentage of consumers are intimidated when selecting wines. They want wines that are easy to identify and that they will enjoy without the need for someone else's help (Hussain, Cholette, and Castaldi 2007). Therefore, to mitigate confusion on what sustainability in the wine industry means, it is important for stakeholders to work collaboratively toward implementing cohesive consumer education programs.

Such programs must help illustrate to the consumer the value of sustainability by explaining its compatibility with traditional winemaking practices. Doing so will facilitate consumer acceptance of such innovation as an improvement rather than a deviation from the coveted norms and traditions of the wine industry. This approach will also aid in creating a positive product image to ease the adoption of such new practices (Garcia, Bardhi, and Friedrich 2007).

Marketing and Competitive Strategies for Sustainability

Marketing Sustainability

Marketing sustainably made wines is still fairly novel for many wine companies. Conventional consumer goods marketing is typically focused either on the customer, competition, or a combination of marketing aspects. The focus of marketing sustainable wines should be more environment-focused to attract a broader segment of the market prone to consume sustainably made wines (Ohmart 2009).

Instead of focusing on one segment of the market, marketing sustainable wines in California, New Zealand, and Australia appears to be broadly aimed. "To increase further, wine must be marketed in a way that will promote everyday drinking among a greater percentage of the population in a socially responsible manner" (Hussain, Cholette, and Castaldi 2007). Therefore, expanding beyond targeting only core wine drinkers is important in also reaching the marginal drinkers, including the younger people in their twenties who may not be as particular, selective, or knowledgeable about wine. While consumers seem to be paying more attention to sustainable marketing messages, many are still confused about what it means to be sustainable, particularly in regard to the wine industry (Ohmart 2009).

Another weakness in marketing sustainability lies in limited capital that most wineries can spend on marketing and hiring trained marketing professionals. It does not appear to be a top priority for small wineries, especially because many of them are not certain where sustainability will go or how it will affect their bottom line. Many wine companies (especially the smaller to midsized firms) feel that marketing is not their strong point, particularly in increasing awareness and demand for their sustainable wines. A winemaker in Australia explained that he was not sure if labeling bottles with indications of sustainability practices will actually help increase his sales (Interview with Alex White at Box Stallion winery, May 29, 2009).

Still, the emerging consensus is that marketing will become an important way to get the sustainability message across to consumers to facilitate demand for their wines. Meanwhile, most winemakers believe that the absolute best way for them to get their message on sustainability across is at the cellar store. Experiential marketing, getting people to experience sustainable wines, appears to most wineries to be the best form of marketing. They believe that those who try their wines become advocates for their brand, their product, and their practices (Box Stallion 2009).

Developing a Cohesive Marketing Message on Sustainability

When developing a marketing message, it is important not to focus on only one aspect of sustainability, but rather expand toward developing a valid and comprehensive depiction of what sustainable wines really are. Collaboration should be encouraged to make a complicated topic simple for the consumer to understand and accept. Marketing sustainable winegrowing should avoid using the term "sustainability" to represent whatever one company wants it to mean. If this term is to survive as a meaningful label in the marketplace, crafting a simple message that is an accurate representation of what it means should become a priority (Ohmart 2009).

Another important aspect is to treat sustainability as a continual improvement process: rather than marketing to customers that they have achieved complete sustainability, the wine industry needs to communicate their story of how they are practicing and advancing sustainably. The wine industry should market sustainability on the basis of facts and resist the temptation to overstate the virtues of one green strategy over another. Giving too much weight to practices that are not entirely sustainable can potentially lead to an industry-wide threat of "greenwashing" (Ohmart 2009). When consumers suspect that a company is overstating their environmentally conscientious acts only to increase profits, they see this as "greenwashing" and do not buy into their message. Such a response could have negative effects industry-wide.

To ensure the validity of marketing messages about sustainability and other "green" practices, some countries have amended existing laws and/or created new laws to encompass "green marketing." For example, Australia's Green Marketing and the Trade Practices Act require claims of being environmentally friendly or sustainable to be true. The law applies "to all forms of marketing, including claims on packaging, labeling, and in advertising and promotion across all mediums . . . any environmental claims you are considering need to be assessed against the requirements of the Act" (Australia 2008).

It seems obvious that the idea of sustainability should be gaining a wider appeal. However, due to differences in regulatory environments, economic conditions, and cultural norms and values, it varies across borders when personal and business priorities interact.

Competitive Strategies

New Zealand

New Zealand has been somewhat forced into the niche and high-quality position due to its inability to reach the economies of scale that neighboring Australia enjoys in the commodity market (Interview with Philip Gregan, New Zealand Winegrowers, May 22, 2009). Because of geography, climate, and size, New Zealand wines have enjoyed being in a niche market for quality wines, which has also been an essential driver of sustainability in this market. Since sustainability practices are typically site-specific, New Zealand is well positioned to continue emphasizing product differentiation. New Zealand has been working hard to exploit their unique climate to produce distinctive wines, and their wine industry demand across the board exceeds supply.

New Zealand wine has also benefited from its national tourism advertising and promotional campaigns. This "halo effect" from advertisements, which emphasize the purity and freshness of its landscape, has helped position New Zealand wine in a way that correlates well with sustainability. Their better wines have been promoted with better marketing, especially with the help of the NZWG's Association (At the Sweet Spot 2008). Of the 70 percent of New Zealand's wine producers who are members of Sustainable Winegrowers New Zealand (SWGNZ), 15 accredited wineries displayed the SWGNZ logo on their labels for their 2007 and 2008 vintages. This right is only obtained when the wine has been 100 percent produced from a winery and vineyard that have been certified as sustainable (New Zealand Wine 2008).

Many wine companies in New Zealand, however, do not perceive marketing their wines as sustainable ones as a high-priority activity. They could use resources to market their wines emphasizing sustainability, but they would rather let the wine speak for itself. As Clark and Yukich of Villa Maria Estate explain, they are "not in the beverage industry, they are in the business of wine where attitudes, human touch, philosophy, and mission" matter most. Leading wine producers such as Villa Maria commit to quality by employing the best people, using the best equipment, adopting the most efficient practices, and therefore do very little (if any) marketing (Interview with Ian Clark and Fabian Yukich, May 21, 2009, at Villa Maria winery). Philip Gregan, CEO of NZW, explained that they have gone back and forth about using sustainability as a marketing and sales pitch, but instead have decided to focus on sustainability as merely a way to continue to produce high-quality wines (Interview with Philip Gregan, New Zealand Winegrowers, May 22, 2009).

Another important reason why New Zealand winemakers do not publicize their sustainable practices much is because they do not want to be viewed as being "weak or soft"; they still try to avoid exploiting their sustainable practices for marketing purposes. This again speaks volumes to the inherent aspects of New Zealand culture, as most if not all of the industry experts interviewed stated that utilizing sustainability as a marketing tool is simply not representative of who they are. Winegrowers in New Zealand see themselves as early adopters and innovators who share a "do what you can with the resources you have" mentality, so sustainable practices are perceived as having been implemented in New Zealand long before the trend began picking up momentum internationally (Clark and Yukich 2009).

The overriding rationale behind adopting sustainable practices in the New Zealand wine industry is clearly not driven by marketing purposes, but rather because it was "simply the right thing to do" (Interview with Philip Gregan, New Zealand Winegrowers, May 22, 2009; interview with R. Ransom of R. Ransom Wines, May 21, 2009; interview with Nicola Belsham, Nicola, May 25, 2009 at Wineseeker and Murdoch James Estate).

Australia

Australian wine competes internationally mainly as commodity wine that has been forced to sell at low prices. The influx of new and inexperienced vintners was largely to blame for the unbalanced supply-demand equation resulting in the downward pressures on Australian wine prices. The lesson learned by the Australian wine market is that firms cannot rely on price alone in their competition. Winemakers can utilize sustainability to compete more effectively in the commodity segment, making wines competitive through both quality and quantity.

In Australia, the industry's approach to sustainable farming may help to drive regional branding and differentiation. However, the challenge will be in cultivating the grapes that are best suited to be farmed sustainably in Australia's various regions, then building public awareness in order to sell them (Cartelle 2008). Australia needs to shy away from the one brand image and shift its marketing focus toward highlighting differences among various regions (Foley 2009), which can potentially be achieved through sustainability practices.

California

In California, economies of scale and scope in marketing offer a way for wine producers to take advantage of America's large population with high purchasing power (Hussain, Cholette, and Castaldi 2007). Despite efforts by certain associations to do a better job marketing and promoting California's sustainable wines, some believe that sustainability practices in farming and producing wines is still mostly "under the radar" (Reign of Terroir 2009). Those in the California wine grape industry believe that their commitment to sustainable agriculture will eventually help them in successfully competing in the global market (Cline 2003).

With the help of CAWG, wine grape growers strive to prove that California produces some of the "best-tasting and highest quality wine grapes in the world." California boasts an unrivaled diversity of climates and soils, making California a leader in sustainable wine grape growing practices and innovation (CAWG 2007). The CAWG facilitates a national California wine country marketing campaign that includes promotional events, public relations, and consumer education with a strong emphasis on sustainable growing practices. Apart from traditional marketing, efforts to market sustainable wines via online platforms include wine association Web sites, wine company Web sites, interactive sites, and more recently wine blogs and social media (Wine-U.S. 2009). In 2007, CAWG implemented a Web site (http://onenationundervines.com) where wine lovers can view a wine firm's commitment to quality, innovation, and sustainable practices. It also provides a way for visitors to explore California's vast wine country (CAWG 2007).

Motivations and Business Logic for Sustainability

Demand for Sustainable Wine

There seems to be a rising consensus that sustainability practices are an improvement upon traditional winemaking techniques. However, whether the consumer demand for sustainable wines will continue to gain further momentum is unclear. About 85 percent of winemakers estimate that there will be an increase in demand for "ecologically and climate friendly produced wine in the future" (Sustainability in the Wine Industry 2009). Thus far, the increased demand for sustainable wine largely results directly from growing consumer awareness of pressing environmental issues.

Wine companies driven by excellence usually hold higher expectations of themselves against the market and customer demand (Gabzdylova, Raffensperger, and Castka 2009). Wine producers in the United States have been responding to environmental awareness among consumers by producing wines in a more environmentally conscious manner (Wine-U.S. 2009). According to Karen Ross, "the California wine community's environmental commitment also contributes to the state's appeal as a great place to live, work, and visit, and resonates with consumers who desire high quality wines produced in a sustainable manner" (California Winegrowing Report 2006).

Most large and some medium-sized companies in New Zealand recognize the importance of sustainable practices, not only for meeting consumer demand, but also for reputation and image. Thus, these firms have focused on enhancing their marketing strengths both domestically and overseas. These efforts often translate into a pull strategy for sustainability in order to meet growing global market demands—when foreign markets begin demanding that exporting wine producers comply with stricter environmental laws, this demand often pulls the wine industry to adopt more sustainable practices. Simply put, marketplace expectation, mainly from overseas customers, drives the demand for sustainability practices. Therefore, sustainable winemaking calls for a proactive approach that (1) brings a good reputation and (2) facilitates meeting such demands with a greater ability to attract suppliers and partners (Gabzdylova, Raffensperger, and Castka 2009).

Financial Implications and Social Return on Investment

There is a growing awareness of the Social Return on Investment (SROI) in the context of sustainability in the wine industry. The SROI concept builds upon cost-benefit analysis, but differs in that it is explicitly designed to inform the practical decision-making of enterprise managers and investors focused on optimizing their social and environmental impacts (Lawlor, Neitzert, and Nicholls 2008). Frog's Leap Winery, among the first wineries in the United States to go green, explains that their motivation was simply because "it just makes sense; financially and physically, it makes sense" (Beer 2009). Since sustainability also means sustaining the livelihood of the business, the additional expenses and constraints that can affect business performance when adopting sustainability must also be considered.

Many wine companies tend to share the belief that green practices are "integral to business because they help run a more efficient operation" (Clark 2009) along with producing a better product that meets the needs of the consumer and the environment. "The first 5 years of operating a green vineyard is more expensive, but after that hump, as long as the vineyard lives, you are operating at an efficiency of 95 percent less input than a conventionally farmed vineyard" (Beer 2009). Frog's Leap Winery invested $600,000 in solar panels that helped them save $50,000 per month in electricity—while also reducing CO_2 emission (Beer 2009). Within one year, the savings realized will cover the cost of the system itself, and the vineyard will not have to spend any more money on electricity for the rest of the 30-years life of the system under the current photovoltaic technology.

The attempts to capture not only the financial value, but also the social and environmental value created by the investments in sustainable practices are also integrated into the "Triple Bottom Line" concept. It captures an expanded spectrum of values and criteria for measuring organizational success: economic, ecological, and social (Brown, Dillard, and Marshall 2006). Government support for sustainable practices may also be a significant motivator in adopting sustainable practices. As part of California Governor's Million Solar Roofs Initiative, California's Public Utilities Commission has mandated that PG&E give credits to customers who feed solar power back into the grid (Clark 2009). These credits, combined with state and federal tax credits, make solar energy more affordable for widespread use. For example, Honig Vineyard and Winery's solar panel system only ended up costing them a third of the total $1.2 million.

New Zealand's Ministry of Agriculture and Farming (MAF) established a Sustainable Farming Fund intended to support projects related to sustainability in farming, including the development of sustainable practices in the wine industry (Improving Energy Use 2009). In one example, MAF funded NZ Winegrowers to identify key areas for improving energy-efficiency in the wine industry, and to establish a benchmark and opportunities for the industry. MAF also offers direct funding to farmers or winery owners wishing to pursue projects aimed at improving sustainability practices. Although most wineries do not receive direct financial benefits from the government, the findings and lessons learned from the projects funded by MAF are published and made available to the industry. Improving the efficiency of their operation and engaging in activities that are "environmentally sound, socially equitable, and also economically feasible" (Brown, Dillard, and Marshall 2006) resonates well with the motivations toward sustainability in the wine industry.

Personal and Ethical Motivations

In New Zealand, sustainability is simply a way of life. Almost everyone interviewed said "it is the right thing to do" rather than being driven by financial benefits or incentives. Villa Maria, one of the largest wineries in

New Zealand, is among the first wineries initiating sustainable practices. "We do things with a '50 years view' and work back from there" (Clark and Yukich 2009). Although not yet fully sustainable, Villa Maria is consistently progressing toward perfect sustainability. They have invested millions of dollars in upgrading their facilities to be more sustainable, and this is primarily because they firmly believe that it is the right thing to do—leave the land as fertile and thriving for future generations while acting responsibly for the natural environment (Clark and Yukich 2009).

Elgo Estate, a smaller boutique winery outside Melbourne, Australia, is fully sustainable and even has negative carbon footprint. According to the winery owner, being sustainable does not cost anything if it is done properly. In fact, it is even more cost-effective. They are proud of their accomplishment in reducing their carbon footprint, and their motivation is solely to protect the land for future generations. "Just as the wines of Elgo Estate are nurtured to reflect the special qualities of this region, it is our responsibility to preserve this beautiful environment" (Interview with Grant Taresch, Elgo Estate winery, May 28, 2009).

In California, the CSWA Chairman sums up the goal of the organization, "to pass the farm on in a better form than it was passed on to me so it can survive the next four generations" (Hoxsey 2006). A recent survey among about 2,000 respondents in the California Wine Industry shows that the top motivations for pursuing sustainable certifications included environmental sustainability, improvement of grape quality, future business viability, and soil quality. The top barriers to obtaining certifications included little financial benefit, unfamiliarity with sustainable practices, and limited market demand" (Anvery et al. 2009).

Conclusion

Grape growers and winemakers consider the land as their livelihood, and have a long history of caring for their land. Although most wineries are unique in their sustainable practices, there are general sustainability trends and goals emerging industry-wide. Consumer demands for sustainable products combined with the changing global economy are driving many wineries toward sustainability. While the actual nonmonetary outcomes of sustainable practices are often hard to quantify, the importance of these outcomes in the long run trumps the short-term financial gains.

Organizations may differ in motivations toward sustainable practices but there is a consistent theme of personal values and a consciousness about the environmental impact that the wine industry makes throughout New Zealand, Australia, and California. Interviews and extensive literature point to an emerging trend—companies receiving organized support from government agencies, winery associations, or simply a sharing of ideas tend to adopt sustainability more readily. This adoption could be in the form of using lighter weight bottles, recycling their own water, or conserving their energy. In each instance, it is clear that the wine industry in New Zealand, Australia, and California is making a significant effort to reduce their environmental impact and preserve their land for future generations.

Acknowledgment

The original study for this chapter was conducted in 2008–2009 by a group of MBA students from Loyola Marymount University, Los Angeles, California. The research culminated with a trip in May-June of 2009 to New Zealand and Australia, including in-depth field interviews with wineries and other organizations and experts in the industry. The group included: Cynthia Aguirre, Frederico Bryansen, Daniella Dorner, Regan Gosnell, Janelle Stack, Laura Vawter, Ryan Wesemann, Anthony Wijaya, and Anatoly Zhuplev (group faculty advisor). The authors express their appreciation for the group members' contribution to the study.

References

About biodynamic agriculture. 2009. *Demeter International.* March 14. http://demeter.net (accessed February 8, 2010).

Anvery, Y., S. Goel, S. Hareesh, J. Hogan, A. Menchaca, and R. Ramirez. 2009. Sustainable practices in the California wine industry: Analyzing the motivations of winemakers and grape growers. UCLA Institute of Environment. http://www.ioe.ucla.edu/academics/article.asp?parentID=2666#9B (accessed April 26, 2010).

At the sweet spot: Wine in New Zealand. 2008. *Economist* 386 (8573, March 29), 85.

Australia. Green marketing and the Trade Practices Act. 2008. Canberra: Australian Competition and Consumer Commission. http://www.accc.gov.au/content/item.phtml?itemId=815763&nodeId=69646a6d15e7958a41b40ab5848c6968&fn=Green%20marketing%20and%20the%20Trade%20Prac (accessed April 26, 2010).

Aylward, D., and M. Clements. 2008. Crafting a local-global nexus in the Australian wine industry. *Journal of Enterprising Communities: People and Places in the Global Economy* 2 (1): 73–87.

Barker, J., N. Lewis, and N. Moran. 2001. Reregulation and the development of the New Zealand wine Industry. *Journal of Wine Research* 12 (3): 199–222.

Beer, J. 2009. Going green inside the wine business. Frog's Leap Winery video interview. http://www.bnet.com/2422-19726_23-210913.html (accessed April 26, 2010).

Brown, D., J. Dillard, and R. S. Marshall. 2006. Triple bottom line: A business metaphor for a social construct. March 3. Portland State University, School of Business Administration. http://www.recercat.net/bitstream/2072/2223/1/UABDT06-2.pdf (accessed April 26, 2010).

Cahill, K. N., and C. B. Field. 2008. Future of the wine industry climate change science. *Practical Winery and Vineyard.* March/April. http://www.practicalwinery.com/marapr08/page1.htm (accessed April 26, 2010).

California winegrowing report reflects increased rise in sustainable practices. 2006. *GreenBiz.com.* December 7. http://www.greenbiz.com/news/2006/12/07/california-winegrowing-report-reflects-increased-rise-sustainable-practices (accessed April 26, 2010).

Carbon Footprint—Home of carbon management. 2010. http://www.carbonfootprint.com (accessed April 26, 2010).

Carbon Trust. 2010. The Carbon Trust. http://www.carbontrust.co.uk/cut-carbon-reduce-costs/calculate/carbon-footprinting/pages/carbon-footprinting.aspx (accessed July 20, 2010).

Cartelle, K. 2008. A review of the Australian drinks industry: Management briefing; Special focus: The Australian wine industry. *Just-Drinks. International News Services.* July. www.internationalnewsservices.com (accessed March 25, 2010).

CAWG initiates California wine marketing campaign. 2007. *Western Farm Press.* June 16. http://westernfarmpress.com/mag/farming_cawg_initiates_california/ (accessed April 26, 2010).

Clark, A. 2009. Green wineries embrace innovation from the fields to the bottle. *GreenBiz.com.* http://www.greenbiz.com/node/26605 (accessed April 26, 2010).

Clark, I., and F. Yukich. 2009. Villa Maria's sustainability efforts. Interview, May 21.

Cline, H. 2003. Sustainable movement growing in California's wine grapes. *Western Farm Press.* March 15. http://westernfarmpress.com/mag/farming_sustainable_movement_growing/ (accessed April 26, 2010).

Cone, T. 2010. Wine sustainability ratings on the way. *The Modesto Bee.* January 16. http://www.modbee.com/2010/01/15/1009909/wine-sustainability-ratings-on.html (accessed April 26, 2010).

Foley, M. 2009. For Australian winemakers, more turns out to be less. *New York Times.* July 3. http://www.nytimes.com/2009/07/04/business/global/04wine.html (accessed April 26, 2010).

Gabzdylova, B., J. F. Raffensperger, and P. Castka. 2009. Sustainability in the New Zealand wine industry: Drivers, stakeholders, and practices. *Journal of Cleaner Production* 17 (11): 992–998.

Garcia, R., F. Bardhi, and C. Friedrich. 2007. Overcoming consumer resistance to innovation. *MIT Sloan Management Review* 48 (4): 82–88.

Global wine industry to tackle carbon footprinting. 2008. *ClimateBiz.com.* January 28. http://www.greenbiz.com/news/2008/01/28/global-wine-industry-tackle-carbon-footprinting (accessed April 26, 2010).

Hoxsey, A. 2006. *CAWG News. California Association of Winegrape Growers.* http://www.cawg.org/index.php?option=com_content&task=view&id=197&Itemid=145 (accessed April 26, 2010).

Hussain, M., S. Cholette, and R. M. Castaldi. 2007. An analysis of globalization forces in the wine industry: Implications and recommendations for wineries. *Journal of Global Marketing* 21 (1): 33–47.

Improving energy use in the wine industry: Stage 1 project report. 2009. *Sustainable Winegrowing & NZ Wine.* http://www.nzwine.com/assets/Improving_Energy_use_in_the_wine_industry_stage_1_report.pdf (accessed April 26, 2010).

Jones, G. V., M. A. White, O. R. Cooper, and K. Storchmann. 2005. Climate change and global wine quality. *Climatic Change* 73: 319–343. http://people.whitman.edu/~storchkh/clim.pdf (accessed April 26, 2010).

Lawlor, E., E. Neitzert, and J. Nicholls. 2008. *Measuring value: A guide to social return on investment*, 2nd ed. London: New Economics Foundation. http://www.neweconomics.org/gen/z_sys_PublicationDetail.aspx?PID=241 (accessed April 26, 2010).

Modern viticulture. 2002. Lodi Winegrape Commission. http://www.lodiwine.com/modernviticulture1.shtml (accessed April 26, 2010).

New Zealand wine: Pure discovery. Annual report. 2008. New Zealand Winegrowers Association. Auckland. New Zealand. http://www.nzwine.com/intro/ (accessed February 8, 2010).

New Zealand winegrowers launch international carbon footprint calculator. 2008. *Winebiz.Daily Wine News.* Australia's wine industry portal by Winetitles. May 2. http://www.winebiz.com.au/dwn/details.asp?ID = 1495 (accessed April 26, 2010).

Nigro, D. 2010. California wine industry establishes new green certification. *Wine Spectator.* http://www.winespectator.com/webfeature/show/id/41550 (accessed April 26, 2010).

Ohmart, C. 2008. Marketing sustainability: Can wine be green without greenwashing? Vineyard view. July 10. http://www.lodiwine.com/Marketing_Sustainability.pdf.

———. 2009. The importance of sustainable winegrowing: Past, present, and future. *Napa Grape Growers Workshop.* Power Point Presentation. April 22. http://www.napagrowers.org/PDF/2009April22/Ohmart.pdf (accessed February 8, 2010).

Potential benefits of sustainable winegrowing practices. 2010. *The Wine Institute.* http://www.wineinstitute.org/initiatives/sustainablewinegrowing/benefits (accessed April 26, 2010).

Organic Wine. 2009. *TM Organics.* March 14, 2009. http://www.tmorganics.com (accessed February 8, 2010).

Osborn, G. 2009. What it means to be GREEN. *Wine.com blog. Wine Notes.* May 22, 2009. http://blog.wine.com/post/What-it-means-to-be-GREEN.aspx (accessed April 26, 2010).

Reign of terroir: Richard Seireeni talks sustainable wine. 2009. May 27. http://reignofterroir.com/2009/05/27/1449/ (accessed April 26, 2010).

Sustainability in the wine industry: On the right track for climate protection. *Climate Partner*. 6 May 2009. Web. 21 July 2009.

Trained sheep can provide vineyard benefits. 2007. University of California. June 1. http://news.ucanr.org/newsstorymain.cfm?story=977 (accessed April 26, 2010).

Wine Institute and international partners to release new Greenhouse Gas Protocol and Accounting Tool. 2008. *The Wine Institute*. January 28. http://www.wineinstitute.org/resources/pressroom/01282008 (accessed April 26, 2010).

Wine–U.S. 2009. *Euromonitor International*. January 25. http://www.euromonitor.com/Wine_in_the_US (accessed via LMU Library March 20, 2010).

CHAPTER 14

New Trends for Sustainable Consumption: The Farmers' Markets as a Business Imperative for the Reeducation of Consumers

Alessio Cavicchi and Benedetto Rocchi

Introduction

Over the past two years, the profound economic crisis that has swept the planet has led to a debate on consumption and the consumption styles of the developed Western countries. Thanks to the media, there are many opportunities for in-depth discussions on the issues of sustainable, ethical, responsible, and critical consumption. While these economic difficulties have severely impacted many businesses, they are also opening up some unexpected roads of opportunity, most of which have been only partially explored. Prominent among these opportunities are those in the domain of consumption.

A recent work, "Measuring and Monitoring Global Consumer Progress Towards Sustainable Consumption" published by GlobeScan on behalf of *National Geographic*, reports a preliminary study in which a small group of experts identified critical behaviors related to consumption. Most of the experts surveyed agree that changes in consumer behaviors are important drivers of sustainability and emphasize the need to reduce the energy intensity of food, mainly by buying local produce but also by buying organic food and avoiding meat.

Questions about the impacts the food supply chain and the associated consumer actions have on the environment, on the landscape, and on the community in general have become pressing and are at the center of the economic debate. Ernest & Young proclaimed Marco Roveda "entrepreneur of the year" in Italy in 1997 in the "Quality of Life" category for founding and managing the "Scaldasole" farm. Roveda, who currently owns Lifegate, an environment-centered firm, stated that "running a business, making it produce a reasonable profit, while framing strategy in terms of both people and planetary welfare is possible. We're all in this together and making it work well is a matter of personal fulfillment, prestige, and commitment to life" (Argante 2008). There is a phrase, possibly a bit overused at this time, to summarize the "mission" implied by that statement: global sustainability. The need for global sustainability is the concept within which the benefits of farmers' markets (FMs) are discussed in this chapter. The aim of the chapter is to underline the role FMs play as a business imperative in reeducating consumers about the value of rural sustainability. An important part of this reeducation process involves helping the urban community familiarize itself with the sights, smells, tastes, and sensations of FMs, as well as the seasonal nature of traditional agricultural production. These are important tasks in avoiding the loss of rural farming traditions and achieving effective protection of rural resources.

This chapter first presents an outline of the global trend of FMs, followed by a discussion of how consumers' attitudes toward FMs are reflecting an increasing interest in local produce. The part of the chapter following that section focuses on the recent proliferation of FMs in one country (Italy) and places the growth of Italian FMs in the context of their history and the peculiarities of the Italian food retail system. The chapter is closed by a final discussion of education issues involved in the diffusion of FMs as an approach to marketing agro-food products consistent with the need for global sustainability.

The Diffusion of FMs: A Global Trend?

FMs around the world are often held up as a key alternative to less sustainable conventional food production systems (Feagan and Morris 2009). FMs are probably the oldest and most common type of direct marketing. Because they are able to bring food producers and consumers together (Feagan, Morris, and Krug 2004; Kirwan 2004), they exemplify an *alternative food network* (Goodman 2004). FMs are the most extreme form of a *short food supply chain* for both time and distance (Renting, Marsden, and Banks 2003). In the past two decades, they have offered a number of consumers their first experience in relocalized and repersonalized forms of exchange.

In the United States, the distance between producer and consumer has grown greatly over the past six decades. The number of family-run grocery shops and street vendors of fruits and vegetables decreased during the postwar period, while mass-market retailers boomed. The revival of FMs began in the early 1970s throughout North America and is ongoing. This reversal has been ascribed to changes in the tastes and lifestyles of American consumers and changes in the national agricultural sector. The number of FMs has almost doubled from 1994 to 2008 (Fig. 14.1). Furthermore, these numbers are likely to underrepresent the real trend; according to Brown (2001), in this FM growth phase it is likely that, not only the number of markets but also the number of stalls in each market would increase, and successful markets would increase their days and hours of operation.

Following the success of FMs in America in the 1990s, several countries (both European and others) have encouraged and supported the establishment of FMs.

In the United Kingdom, the recent rapid increase of FMs started in 1997. The increase in local FMs (from 200 in 2000 to 550 today) has reached an annual revenue of over 300 million Euros, involving about 15 million highly loyal consumers. The number of customers has almost doubled over the past two years.

In Germany, there are more than 5,000 active markets, one of the most important of which is Coburg in Bavaria, founded in January 1992. In France, the number of markets exceeds 500. There are similar rising trends in Canada and Australia.

The growth in FMs is a common part of the evolution of the food-supply chain in the postindustrial phase of society. Interestingly, in China, it is possible to detect a countertrend with respect to the more developed countries' data as China rapidly follows the steps taken by Western countries during the postwar period (Gale 2003). Modern "hypermarkets," convenience stores, and fast food restaurants are taking the place of traditional FMs

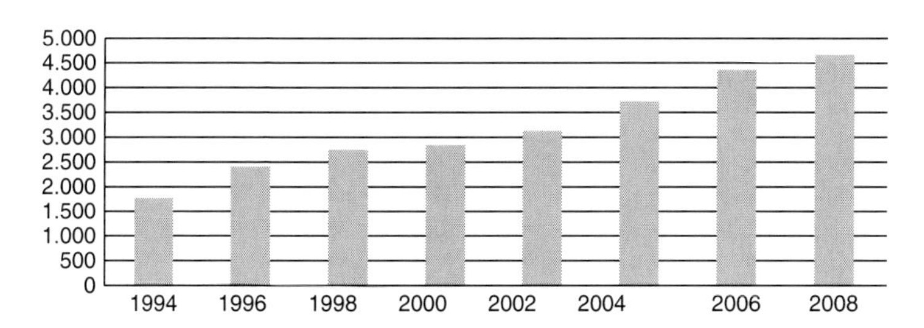

Figure 14.1 Changes in the number of FMs in the United States
Source: USDA – AMS Marketing Services Division

and corner kiosks. Open-air FMs still offer city residents fresh vegetables, fruits, eggs, and meat; but the nature of these traditional food outlets has changed. In addition to their own produce, these vendors now sell products mainly purchased from large wholesale markets. It is obvious that in China the new transportation, storage facilities, and likely increasing use of factory farming techniques that will occur as the modern commercialization of the food sector progresses will cause new sustainability problems.

Toward Sustainability: Consumers' Attitudes and Motivations

No single formula for an FM exists. Markets will differ depending on the context in which they occur and the consumer segment to which they are addressed. However, in many of the international cases examined by Traclò (2009), there are elements that seem to be common to all or most FMs:

- Goods are transported only short distances;
- Ways exist for consumers to verify the quality and origin of food products;
- Activities involving food traditions are included in the markets;
- For the most part, markets are weekly events;
- The clientele is highly loyal; and
- Groups organize cooking lessons and nutritional awareness sessions in the markets or through them.

In recent years, especially in the United States, some research has been conducted with the aim of understanding the motivations, perceptions, and willingness of consumers to shop for and purchase FM products. One example is the study of 336 customers in the San Luis Obispo, California FM (McGarry, Spittler, and Ahern 2005) that noted how quality and value of products are among the most important attributes pursued by customers. Customers valued freshness, taste, convenience, reasonable price, and low environmental impact. Of course, such factors as freshness and environmental impact are easier for consumers to assess for FM products because the possibility of tracing the path of the supply chain is more readily done than with purchases from a supermarket.

Even taking into account differences among regions and the sociocultural backgrounds of the resident populations, the factors highlighted in the research carried out in San Luis Obispo seem to be present in many other countries. International studies suggest that the most important factor leading people to purchase from FMs is not the price, as might be thought given the expansion of the phenomenon during the recent economic crisis, but rather quality. Consumers usually cite "better food quality," "locally produced foods," "higher social interaction," and "learning directly about the vendors and their food production practices" as the principal motivations for buying in FM environments (Gale 1997; Govindasamy et al. 1998; La Trobe 2001; Halweil 2002).

Zepeda (2009) notes female food shoppers are more likely to patronize FMs than males. In her research, education, age, and economic variables such as income level do not appear to affect the probability of purchasing local foods. Instead, attitudes toward fresh and local food are the relevant factors. As a consequence, freshness, environmental concerns, health issues, and the desire to support local farmers represent the right bases to cover in communication with existing customers. Those customers may be especially loyal if there is another adult in the family and if they desire to cook their own meals. Conversely, communication campaigns targeting nonshoppers, such as males and singles with low cooking skills and scant free time, need to be based on fast and easy-to-prepare meals on the basis of fresh ingredients and on convenience, such as the few moments needed to visit a nearby FM. One study found that visits to FMs averaged only 16 minutes (Baber and Frongillo 2003).

Recently, Thilmany, Bond, and Bond (2008) explored consumer behavior and motivations for direct food purchases. They note that the *New Oxford American Dictionary* proclaimed "locavore" word of the year in 2007. They describe locavores as consumers who are aware of the impact of food selection on the environment and who look for locally produced foods and beverages.

Which psychological and sociological mechanisms are related to this rediscovery of local food? Gabbai, Rocchi, and Stefani (2003) explored the psychological and sociological mechanisms related to this rediscovery of local food in a review of the main factors shaping local food choices. They cite the work of Ray (2008) who defines some products as *cultural markers*, suitable for a postmodern consumer who consciously and with an aesthetic sense looks for meanings and ideals in new communities. This perspective aligns with the rising interest in authenticity and in the symbolic role of food mentioned by Bessière (1998), who writes about the *culinary heritage* of a certain area.

Another basic reason for decisions to consume local food is framed in normative and moral terms. Through the selection of local food, it is possible to buy something that will help local and marginal communities become more sustainable and at the same time consume something *pure* and *natural*, as opposed to the "*artificial*" supplied by the modern agri-food industry. Bessière defines this attitude as *gastronomical curiosity*. These perspectives suggest the incorporation of place, culture, and community identity into foods.

FMs in Italy: An Ancient Story

In Italy, the concept of a short food supply chain is only "relatively" new. Great examples of FMs can be found in the history of Italian medieval cities, and in recent times people have been able to purchase fruits and vegetables directly from farmers who come regularly to the city with their farm produce. However, what is different about the Italian FM history is the way the number of FMs has ebbed and flowed over the centuries, how the FMs have been related to other food suppliers, and their considerable resurgence in recent years.

In the first centuries of Roman history, farmers and ranchers came to town to sell their wares directly, but by the time of Imperial Rome, more than 150 independent corporations had come into being. Those corporations managed the food supply chain step-by-step from producers to wholesalers to retailers. Only a few merchants (notably gardeners and fishmongers) worked at the same time as producers and retailers (Traclò 2009).

With the decline of the Empire and the emergence of a feudal economy, the tendency of individuals to produce and process their own raw foods regained prominence, with a resulting decrease in the number of shops in the urban markets. Much of the economic wealth came from the cultivation of fields; the semiwild breeding of oxen, horses, pigs, and sheep; and the exploitation of natural resources in the form of hunting, fishing, and harvesting of wild vegetation (by both legal and illegal means). Those products were normally consumed in the same places where they were produced, by those with a role in producing, harvesting, or distributing them (Traclò 2009). In the manorial world, the few trades that existed took place in the rare markets held weekly, monthly, or even annually (Cipolla 1974).

Between the tenth and the thirteenth centuries, a revival and development of towns was accompanied by the simultaneous growth of agriculture, enabling the feeding and support of growing urban populations. In the twelfth century, ancient cities began to shine once more. Those cities and other recently founded new ones became natural places for periodic markets and fairs. The markets were less spontaneous and random than one might imagine, occurring at specific places (only one place in each city, set aside for particular kinds of exchange), at defined times (every day except public holidays, with different durations according to the times of sunrise and sunset during the year), in specified spaces (precisely delineated market areas with only pedestrian traffic when the market was open), and governed by very specific rules and customs (with special commercial statutes fostering a transparent comparison among the goods for sale on the counters and rules for negotiations between merchants and customers). Street vendors and those in market stalls were compelled to follow the rules specified in the statute establishing the market and by the social control tacitly exercised by the large number of sellers and buyers who flocked there (Traclò 2009).

This rural and urban economic equilibrium did not radically change before the outset of industrialization in the last decades of the nineteenth century. The structure of food supply chains in Italy was deeply changed only in the postwar period. The rise of urban economies fostered by industrialization led to modernization of retailing as well. After pioneering entrepreneurial initiatives at the end of the 1950s, modern retailing sharply increased for two decades. New forms of retailing such as supermarkets were welcomed by consumers as a sign of modernization and also for high-quality products at low prices (Scarpellini 2004).

However, the transition to an "industrial" structure for the food distribution network in Italy is still incomplete. The degree of concentration in food retailing in Italy is still far from the level present in other developed economies. A strong geographical differentiation also exists, with the southern region showing a higher incidence of traditional forms of retailing. Indeed, Italy has experienced a development of the retail sector totally different from the development found in many other countries. The fragmentation that prevails, together with the small size that has characterized the retail sector for many years, has kept the price of goods in general, not merely food, more expensive for consumers. In addition, highly restrictive legislation (Act No. 426 of 1971) that protected small retailers and traders already in the market hindered the development of modern commercial distribution and the entrance of foreign competitors into the Italian market. This phenomenon, singular in the global context, has delayed the unchallenged reign of the modern grocery retailers, at least in rural areas and in southern regions. As a result, traditional food retailing has had the opportunity to be sourced for years from local suppliers that all customers could get to know directly.

The recent history of FMs in Italy has roots that date back to the early 1970s as part of a return-to-the-land movement. That movement seeks production methods not dominated by the industrial agri-food system: methods that will ensure the survival of different forms of agriculture responsive to consumer demand (Rossi, Brunori, and Guidi 2008).

In recent years, the modernization of the retail food system has been associated with an increasing awareness of the problems generated by the distance between food production and consumption. A reform of trade regulation in 1998 (the so-called Bersani Decree) supported further modernization of the retail sector. This measure allowed the entry of foreign retailers into the Italian market through mergers and acquisitions of existing local chains (Cavicchi 2007). This liberalization, however, also allowed the entry of marketing methods in distribution considered "extreme" for Italian shopping culture, such as discounts to attract customers as well as expanding the consumption of food products purchased from progressively more distant suppliers. After initial enthusiasm, there has been some rethinking about the modern distribution network, the organizational structures and approaches that have emerged. Ephemeral, local, and seasonal products have been gaining more and more space, even in the stores of those foreign producers who had managed to penetrate the Italian market.

This gradual change is also due to new styles and habits of sustainable consumption, which are increasingly apparent among all the contradictions that characterize postmodern consumer behavior. An analysis conducted by Federalimentare-Ismea (2004) using 2003 Ismea-ACNielsen data identified two macrotrends typical of an urban area characterized by high availability of income and a hectic lifestyle. Consumers orient their food choices to save time while also simultaneously following food traditions. These two requirements seem to conflict, but industry and food retailers play upon their interaction to differentiate their products in pursuit of their marketing strategies. These dimensions can potentially facilitate the consumption of local products too, provided that the consumer is properly informed.

The negative impact on final product price for consumers generated by the distance between the sites of food production and consumption has often been discussed. The Italian Competition Authority (Antitrust) carried out an investigation (Autorità Garante della Concorrenza e del Mercato 2007) on the distribution of food products, concentrating on the structure and functioning of the produce sector. It concluded that the structure of production and distribution of Italian fruits and vegetables needed to be changed to prevent too many players along the supply chain, whose existence would increase the price to final consumers to an abnormal extent. Shortening the chain of distribution was recognized as one crucial element that can increase the efficiency of the sector and the welfare of final consumers by decreasing prices. Long supply chains often lead vendors to charge prices of up to 300 percent over cost.

A survey carried out by Trendfood Aegis Media, Italy (http://trendfood.it/), showed that pressures from sustainable food lovers are leading mass-market retailers to increase their offerings of traditional, wholesome, and natural food because "the ability to pragmatically embrace the environmental issue can be an important branding lever. . . . " However, those same consumers also appear to be lovers of FMs and support initiatives such as the *KM0* (zero food kilometers) "short supply chain" initiatives promoted by associations such as Slow Food. FMs shorten the food supply chain and emphasize the relationship between producers and consumers (or consumer groups) by reducing the length of the route food takes to market and the number of agents involved. The final distribution area for FMs roughly corresponds to the production area.

In recent years, new regulations have been adopted that are more oriented toward the facilitation of direct marketing by farmers. These regulations are consistent with EU efforts to support rural development that fosters new commercial outlets for local and native products. In 2007, a new impetus was offered by a decree of the Ministry of Agriculture regulating FMs according to the following salient points:

- municipalities can establish or authorize the agricultural markets that meet the standards specified in the decree;
- direct sales of agricultural products may be established in public areas, in premises open to the public as well as on private property;
- only farmers operating within the region or in areas defined by local institutions can participate in these forms of distribution—selling products from their own firm or from a company of partner farmers, or from their food processing activities—in compliance with sanitary regulations;
- within these markets, cultural, educational, and demonstration projects related to food and traditional crafts can be undertaken, provided they refer to the same rural area, though farmers can sell their production in tandem with farmers from other areas when there are synergies and authorizations to do so. *Coldiretti* estimated that about 500 FMs were started in Italy under the "Friendly Countryside" program.

Consumer Behavior: Some Evidence in Italian Markets

Italy's strong culinary heritage, bolstered by historical, political, institutional, and cultural factors, supported the resilience of traditional forms of food retailing in cities. Those factors also supported local farmers' direct sales of regional specialities such as wine or olive oil in Tuscany. The ongoing movement toward relocalization of food supply, which got underway in the early 1970s, found a *significant* share of Italian consumers with positive attitudes about this form of distribution. The profusion of FMs in Italy in recent years appears in many cases to be the result of a change in consumer *motivation* rather than a deep change in shopping habits.

In recent years, FMs have allowed many consumers to rediscover genuineness, naturalness, and freshness in food products, providing not only quality products but also a new atmosphere of trust in the producer. Due to salient food-related health problems and scandals in the past decade, such as "mad cow" disease, "avian flu," "Mozzarella di Bufala dioxin," and others, a renewed relationship between sellers and consumers is needed even more than in the past. This need provides new paths for rural development within a sustainability approach. Moreover, FMs can be part of the renewal of cities.

Many FMs that originated in part as a reaction to the decline of rural areas are now becoming a high-growth part of local economies. Data from the *National Observatory* on spending in Italy shows that in 2007, 2.5 billion Euros were spent on the purchase of wine, fruit, vegetables, olive oil, cheeses, and other delicacies from 57,530 registered farms in Italy, with an overall increase of 48 percent since 2001 (*Coldiretti* and Agri 2000). During the year, seven out of ten Italians made purchases at least once directly from farmers with a perceived cost savings of 30 percent, even though these purchases were mainly made because of the quality and freshness of the products (http://www.coldiretti.it/docindex/cncd/informazioni/141_10.htm). In 2009, the Swg-*Coldiretti* survey contained additional information about an increase in the frequency of visits to and purchases from farmers. Almost two out of three consumers have been purchasing from farmers at least occasionally, while more than one out of ten did so on a regular basis. The survey reported that the number of farmers selling directly has risen to 60,700 with a turnover of 2.7 billion Euros. The elements of satisfaction for those who buy are listed as the atmosphere and the personal relationship with the producer, followed by freshness, price, and assortment of products.

For consumers interviewed at the FM in Fermo, Marche Region, (Cavicchi and Donati 2010), quality was defined as freshness. It appears that the most suitable strategies for FMs are those that will increase the possibility for consumers to find authentic and natural foods associated with healthy lifestyles. This message might usefully be spread among younger people in particular, along with the opportunity to shop at times that regular grocery stores are closed to emphasize the greater convenience that FMs can provide.

A survey of over 90 customers in several FMs and shops in Tuscany (Baldeschi 2009) provided evidence that FM shoppers were typically women, 40–50 years old, with a higher than average level of education and financially well off. Shoppers at FMs were found to be of two sorts. The first group is comprised of people with high levels of income and education who regularly buy food that is certifiably of high quality (indicated by such things as origin or organic methods of production) and who are less concerned about price. It is not direct contact with producers that is valued but rather confidence about product quality and confidence in the trustworthiness of FMs as institutions. Some regions such as Tuscany have developed policies to secure confidence in the trustworthiness of their FMs.

The second group is comprised of consumers who are less well off than average. They were motivated to shop at FMs for price and trust reasons. These consumers were observed to purchase a wider range of products and to spend more on each visit to FMs.

The FM experience might well vary across nations. Customers in Italy may be usual in the extent to which they shop for social reasons and to support environmental sustainability, along with an interest in experiencing a traditional approach to shopping.

Concluding Remarks: New Challenges and Opportunities to Reeducate Consumers

As recently stated by Sexton (2009): "When the desired quality characteristics of the food products themselves are considered, opportunities are created for well-positioned small firms to exploit market niches." *Locavores*, with their desire to consume only food products produced near their home are attractive customers for FMs and the shorter supply chains the FMs create. Farmers are motivated to move to a short supply chain by higher remuneration than can be obtained by selling to the agro-food industry. Consumers find greater convenience, enhanced transparency, and more confidence in information provided by the seller, as well as fresher and healthier food. This virtuous process can stimulate rural development in marginal areas and benefit the community by reducing transport with a resulting decrease of environmental impact.

A key argument for this new trend of sustainable consumption is proposed by Eden, Bear, and Walker (2008) who observed that consumers in developed and developing economies are increasingly disconnected or distanced from producers, and that this distancing has allowed producers and retailers to supply food that is increasingly dangerous to human health, the environment, and development. They observe that the largest impacts are on developing countries. Reversing this distancing is not an easy task. It requires reeducating consumers about the food values and traditions of the places where they live.

According to Thilmany, Bond, and Bond (2008), even if the developed countries are experiencing an increasing demand for local foods, there is still need to study the motivations of consumers and their willingness to pay for FMs' products, the role of direct markets in consumer shopping choices, and the buying profiles of consumers who frequent FMs.

As Cavicchi and Corsi (2010) state, the consumers' desires to support, protect, and assert their identities may be expressed by the purchase of local products. Therefore, marketers should articulate how the purchase of a local product enables consumers to strengthen their sense of belonging to their region while supporting the local economy. An appropriate educational process aimed at all supply chain actors, from farmers to consumers, can promote the recovery of historic values and communicate traditions that are at risk of being lost. A research and training process aimed at gaining and diffusing knowledge about both intrinsic and extrinsic quality attributes of local food products and the impact on food purchasing decisions of complex historical patterns, social habits, sensory characteristics, and more general quality cues is needed.

Marketers of regional and local products must differentiate their communication strategies: targeting quality-conscious consumers, who appreciate naturalness and craftsmanship and who have a strong sense of belonging to the region, while communicating the virtues related to the convenience of easy meals prepared with fresh foods to attract new potential consumers looking for time-saving purchases, and increasing pricing transparency to motivate consumers looking for less costly ways to get quality foods (Baldeschi 2009). The work of Zepeda (2009) is consistent with this perspective, showing that when FMs are communicating with potential new customers, they need to emphasize the convenience factor and when communicating with existing customers, they should continue to emphasize freshness, positive environmental impacts, and health contributions.

In Italy, the incomplete modernization of food retailing systems allowed the continuation of traditional retail point of sales, such as local weekly fairs and direct sales from wineries, and the FM movement is building on solid consumption habits and generally favorable attitudes toward the "food mile," even in cities. This fact gives producers the chance to base their businesses on traditional, sustainable consumption habits without needing to rely only on trendy or ideological interests in adopting sustainability-oriented habits. Despite these favorable conditions in Italy, an educational effort is still necessary to strengthen the shift toward more sustainable consumption habits that FMs offer.

These education/communication strategies should be pursued through the joint work of economic actors and public institutions interested in developing an area's reputation. To enhance the value of rural capital, educational efforts should be addressed not only to tourists visiting the rural environment, but also to city dwellers, educating them about the sights, smells, tastes, and sensations of the fresh foods that FMs offer, as well as the seasonal nature of traditional agricultural production. These educational endeavors are crucial in avoiding the loss of local traditions and achieving an effective promotion of rural capital through the FMs that grow and maintain that capital.

References

Argante, E. 2008. *Marco Roveda, l'Ecobusiness ci salverà* [*Marco Roveda: Ecobusiness will save us.*]. Rome: Salerno Editrice.

Autorità Garante della Concorrenza e del Mercato. 2007. *Indagine Conoscitiva sulla Distribuzione Agroalimentare*, IC28. http://www.agcm.it/AGCM_ITA/DSAP/DSAP_IC.NSF/8c140a0d4d64cba941256262003d5c11/8d9113caebab738cc12572fb003ce5d3/$FILE/IC28.pdf (accessed April 24, 2010).

Baldeschi, M. 2009. *I consumatori e i mercati dei produttori: un'indagine esplorativa* [*Consumers and producers' market: Fact-finding*]. Unpublished master thesis, College of Agriculture, University of Florence.

Baber, L. M., and E. A. Frongillo. 2003. Family and seller interactions in farmers' markets in upstate New York. *American Journal of Alternative Agriculture* 18: 87–94.

Bessière, J. 1998. Local development and heritage: Traditional food and cuisine as tourist attractions in rural areas. *Sociologia Ruralis* 38(1): 21–34.

Brown, A. 2001. Counting farmers markets. *Geographical Review* 91(4): 655–674.

Cavicchi, A. 2007. La distribuzione alimentare [Food Distribution]. In *9° Rapporto Economia e politiche rurali in Toscana*, ed. IRPET. Rome: Agrisole—Il sole 24 ore.

Cavicchi, A., and A. Corsi. 2010. Consumer values and the choice of specialty foods: The case of the Oliva Ascolana del Piceno (protected designation of origin). In *Market orientation transforming food and agribusiness around the customer*, eds. A. Lindgreen, M. Hingley, D. Harness, and P. Custance. Aldershot: Gower Publishing.

Cavicchi, A., and C. Donati. 2010. Vendita diretta, "KM 0" e nuovi stili di consumo: il caso del Farmer market di Fermo [KM 0 and new styles of consumption: The case of Fermo farmers' market]. In *Annali della facoltà di Scienze della Formazione*. Macerata: CEUM.

Cipolla, C. M. 1974. *Storia economica dell'Europa Pre-industriale*. Bologna: Il Mulino.

Coldiretti. http://www.coldiretti.it/docindex/cncd/informazioni/141_10.htm (accessed August 3, 2010).

Eden, S., C. Bear, and G. Walker. 2008. Understanding and (dis)trusting food assurance schemes: Consumer confidence and the "knowledge fix." *Journal of Rural Studies* 24(1): 1–14.

Feagan, R. B., and D. Morris. 2009. Consumer quest for embeddedness: A case study of the Brantford Farmers' Market. *International Journal of Consumer Studies* 33: 235–243.

Feagan, R. B., D. Morris, and K. Krug. 2004. Niagara region farmers' markets: Local food systems and sustainability considerations. *Local Environment* 9: 235–254.

Federalimentare-Ismea. 2004. *L'industria agroalimentare in Italia* [*The food industry in Italy*]. *2° Rapporto Federalimentare-Ismea*. Assemblea Federalimentare, Parma, 6 maggio 2004. www.federalimentare.it/Documenti/FEDERALIMENTARE-ISMEA.pdf (accessed April 24, 2010).

Gabbai, M., B. Rocchi, and G. Stefani. 2003. Pratiche alimentari e prodotti tipici: un'indagine qualitativa sui consumatori [Eating practices and typical products: A qualitative survey of consumers]. *Rivista di Economia Agraria* 58(4): 510–552.

Gale, F. 1997. Direct farm marketing as a rural development tool. *Rural Development Perspectives* 12: 19–25.

Gale, F. H. 2003. China's growing affluence: How food markets are responding. *Amber Waves* 1(3).

GlobeScan. 2007. *Measuring and monitoring global consumer progress towards sustainable consumption*. National Geographic Society.

Goodman, D. 2004. Rural Europe redux? Reflections on alternative agro-food networks and paradigm change. *Sociologia Ruralis* 44(1): 3–16.

Govindasamy, R., M. Zuerbriggen, J. Italia, A. Adelaja, P. Nitzsche, and R. VanVranken. 1998. *Farmers markets: Consumers trends, preferences, and characteristics*. The State University of New Jersey, NJ: Department of Agricultural, Food and Resource Economics, Rutgers Cooperative Extension.

Halweil, B. 2002. *Home grown: The case for local food in a global market*. Worldwatch Institute, World Watch Paper 163, Washington, D.C.

Kirwan, J. 2004. Alternative strategies in the UK agro-food system: Interrogating the alterity of farmers' markets. *Sociologia Ruralis* 44: 395–415.

La Trobe, H. 2001. Farmers' markets: Consuming local rural produce. *International Journal of Consumer Studies* 25: 181–192.

McGarry, M., A. Spittler, and J. Ahern. 2005. A profile of farmers' market consumers and the perceived advantages of produce sold at farmers' markets. *Journal of Food Distribution Research* 36(1): 192–201.

Ray, C. 1998. Culture, intellectual property and territorial rural development. *Sociologia Ruralis* 38: 3–20.

Renting, H., T. Marsden, and J. Banks. 2003. Understanding alternative food networks: Exploring the role of short food supply chains in rural development. *Environment and Planning A* 35: 393–411.

Rossi, A., G. Brunori, and F. Guidi. 2008. I mercati contadini: un'esperienza di innovazione di fronte ai dilemmi della crescita [Farmers' markets: An innovative experience facing growth dilemmas]. *Rivista di Diritto Alimentare* 2(3): 21–26.

Scarpellini, E. 2004. Shopping American style: The arrival of the supermarket in the postwar Italy. *Enterprise and Society* 5: 625–668.

Sexton, R. J. 2009. Forces shaping world food markets and the role of dominant food retailers. *Paper presented at the XLVI meeting of Societá Italiana Di Economia Agraria,* September 16–19, 2009, Piacenza (Italy). http://ilo.unimol.it/sidea/

images/upload/convegno_2009/plenarie/relazione%20plenaria_sexton.pdf (accessed April 26, 2010).

Thilmany, D., C. A. Bond, and J. K. Bond. 2008. Going local: Exploring consumer behavior and motivations for direct food purchases. *American Journal of Agricultural Economics* 90(5): 1303–1309.

Traclò, F. 2009. *Mercatipico.* http://www.restipica.net/index.php?option=com_phocadownload&view=category&id=5%3Aarticoli-di-res-tipica&download=113%3Avolume-mercatipico&Itemid=11&lang=en (accessed March 10 2010).

Trendfood. http://trendfood.it/ (accessed March 2010).

Zepeda, L. 2009.Which little piggy goes to market? Characteristics of U.S. farmers' market shoppers. *International Journal of Consumer Studies* 33: 250–257.

CHAPTER 15

Participation of Agribusiness Stakeholders in Global Sustainability Questions: The Case of Climate Change and Bioenergy in Brazil

Denise Barros de Azevedo, Eugênio Ávila Pedrozo, and Guilherme Cunha Malafaia

Introduction

Stakeholders are increasingly leading firms into the consideration of sustainability issues with local, national, and global ramifications. Managers in many nations are increasingly seeing global sustainability issues as legitimate concerns for their businesses. This chapter accepts behavior at the level of individual managers and individual stakeholders as critical in addressing global climate change and focuses on the need for individuals and institutions to engage in dialogues about climate change and agribusiness—dialogues involving stakeholders from different areas of knowledge and with different interests.

The need for such dialogues is particularly pressing because global industrial development has been and still is heavily dependent on energy sources, and agribusiness-based biofuels are seen by many as a vehicle for dealing, at least in part, with the myriad problems created by continuing dependence on nonrenewable and environmentally destructive fossil fuels. However, at the same time, it is also clear that the recent increased reliance on biofuels to replace fossil fuels has carried with it its own set of problems and conundrums.

The large role of nonrenewable sources in the world energy supply challenges societies to come up with alternative energy sources. The search for bioenergy sources and supplies has become part of organizational and national strategies for economic growth. While biofuels do offer opportunities for reducing the world's reliance on fossil fuels, they do raise food supply concerns. The impact on food security varies in different places and times as market forces evolve and related technologies are developed. Political choices at national and international levels play an important determinant of the role various forms of bioenergy will play. An inclusive dialogue among the bioenergy stakeholders and organizations considering using them is one possible way to begin developing solutions to disagreements and for balancing competing interests.

Climate Change, Fossil Fuels, and Bioenergy

Climate change is a salient topic in current scientific, industrial, and agricultural discourse, as well as in economic and political forums and civil society. Addressing global sustainability issues in corporate strategic development is a business imperative. Concerns arise from poor use of soil, natural resources, minerals, water resources, the atmosphere, and forests. Serious environment problems are occurring more often including floods, droughts, tornados, severe hurricanes, rising sea level, among others (Metz 2010; FAO—Food and Agriculture Organization 2008).

The Impacts of Fossil Fuels

The literature puts forth several phenomena that are directly related to the intensification of the greenhouse effect resulting from the burning of fossil fuels. Some American scholars have explained that forests and oceans, which have long functioned as drains or deposits of carbon dioxide (CO_2) from excesses in the atmosphere, are losing this capability due to the saturation of the systems, which might be one of the causes of the observed abnormal increases of the concentrations of CO_2 (Metz 2010). Because of this, it is feared that the greenhouse effect will intensify with catastrophic alterations resulting from global warming such as rising sea levels and a worsening of droughts and storms.

Facing these issues, the search for renewable energy, in this case bioenergy, becomes a strategy for the advancement of organizations, society, and stakeholders. The huge use of nonrenewable sources in the world energy supply gives society the challenge of focusing on the search for alternative sources of energy (Smith and Taylor 2008). The world is increasingly aware of the negative impact of fossil fuels on climate (including more severe droughts, floods, hurricanes, and tsunamis) and the opinion of leading scientists that climactic extremes are occurring more frequently and harshly.

Business Responses to Climate Change

Although many organizations consider green business practices to be too overly constraining to adopt, other organizations see business opportunities and competitive advantages lurking in actions associated with promoting sustainability in corporate undertakings, including products and services (Porter and van der Linde 1995), especially because of the emergence of new market niches involving sustainability.

A burgeoning demand for environmentally helpful products has led to the development of new product standards and new business opportunities, sparking environmental legislation aimed at changing current industrial and agricultural processes. Integration of concern for the environment deeper into socioeconomic structures fostering a harmony between economic growth and environmental preservation is moving from seeming to be impractical to seeming to be possible.

Thus, pressures generated by society and the market are leading to the development of new actions involving, for example, the use of CDM (clean development mechanisms), the handling of natural resources, organic production, environmental certification, sustainable reforestation, carbonic gas reduction procedures (Kyoto Protocol) and, most importantly, empowerment of stakeholders. Indeed, the influence of the stakeholders of organizations is now a prominent topic for researchers and society.

The Role of Agribusiness and Bioenergy

A paradoxical situation is arising for agribusiness where many nations currently exporting agricultural products will find their agricultural production systems disrupted by climate changes such as different precipitation levels and lower or changing biodiversity. On the other hand, leading agribusiness nations will find new global business opportunities in such endeavors as reforestation, biofuel production, carbon capture, and natural resource nurturing.

Brazil and the Bioenergy Situation

The situation in Brazil is different from the situation in the United States and Europe (notably Germany) due to the option to utilize sugar cane and the greater availability of land (FAO 2008a). Ethanol originating from sugar cane reflects large recent harvest in Brazil of around 632 million tons with about 55 percent destined for the production of biofuels. Yet, this production level comes from but 0.4 percent of the area with grains plantations in Brazil. However, 27 percent of the increase in sugar cane production occurred in areas that were previously planted with soy, corn, coffee, and oranges.

The Bioenergy and Food Trade-Off: An Important Case for Global Sustainability

The Food and Agriculture Organization of the United Nations (FAO) (2008) points out that increases in grain prices generate social tensions and acts of violence in poorer countries (FAO 2008). World grain consumption will increase by 2 percent to 2.1 billion tons, while the stocks of 143 million tons are in the lowest levels of the past 25 years (*Agroanalysis* 2007: p. 29). One of the versions attributes this to ethanol, which would be responsible for 75 percent of the growth. In 2008, the global cost of the importation of food was tracking to reach USD 1.035 billion, 26 percent more than 2007.

Also according to the FAO (2008), several factors amplify the problems with the production of bioenergy, such as increasing food demand in countries such as China and India; the steep ramping up of biofuel production involving the use of around a hundred million tons of grains for ethanol from corn, concomitant with American and European Union (EU) subsidies; commodities speculation; increasing oil, freight, and fertilizer prices; and losses of crops due to climatic factors, for example in Australia.

The most pressing question is whether biomass production will dislocate an undue amount of other critical productive resources (land, work, and capital) from food production to the cultivation of grains destined to produce fuels. Corn is the focal crop in this regard, and the United States is prominent as the world's largest producer and exporter of corn. Worldwide there has been an increase of 224 percent in the price of corn from 2005 to 2008 (FAO 2008).

In face of this scenario as described by FAO (2008a), it is clear that world production and supply of food is in crisis, with a high inflation of food prices associated with the negative factors derivative from climate changes aggravating this problem.

This elevation of prices dramatically affects the 2.5 billion people who live with or even less than USD 2 a day. It is important to emphasize that Brazil is able to face the agricultural price crisis because of the presence of a vigorous family farming sector that produces 70 percent of the food consumed by Brazilians. Since 2003, Brazilians have developed a strategy to strengthen their agriculture, with public credit policies, agricultural insurance, technical assistance, and rural extension (Cerri 2009).

The use of biofuel incentives concerns the determination of whether there are effects on the dual markets of bioenergy and the food sector. That is, the question remains whether the expansion of biomass use for energy bolstered sustainable food production or the opposite. Preliminary results confirm a positive effect. Environmental and social issues, particularly the generation of jobs, have been shown to influence the synergetic expansion of these sectors. These synergies create "spatial" economies (wide area of coverage) as well as a "scale" economy (economies of scale), and consequently result in "added value" to the final product and to a higher rate of social return (Yeganiantz et al. 1984).

Bioenergy offers opportunities and risks for the securing of required food supplies. Impacts on food security can vary in space and time depending on the evolution of the market forces and technological developments, both influenced by political choices at national and international levels. It is necessary to develop an analytical structure that considers the diversity of factors contributing to the situation and the specific needs of countries (Yeganiantz et al. 2007).

The comprehension of the use of dialogue among stakeholders can be thought to portend a better alternative for attaining solutions to the related conflicts to bring us to global sustainability of the natural environment and prospering organizations as a business imperative.

A review of the literature revealed important questions concerning the energy crisis in the world, including the dilemma of the choice between producing food and producing bioenergy contextualized by the use of stakeholder dialogue to present new alternatives for addressing climate change.

Food and bioenergy do compete. For example, the Co-op Insurance Society report indicates that 9 percent of the world's agricultural lands may be required to offset the agricultural production that supplies world oil transportation. That is, the production of bioenergy could lead to a decrease in the available land for food production, and that it would be an especial problem in countries where food is already scarce (Griffin 2007).

According to Griffin (2007), this could intensify the problem due to the fact that the population in the world will probably increase by about 1 billion people by 2015 and, in 20 years it will be necessary to increase food production by 50 percent to feed the growing world population.

In this calculus of agriculture for food production versus for energy production is the impact on food prices and the possibility that markets will find food price rising above what many people can afford to pay for subsistence requirements. Decades of experience demonstrate that food "availability" does automatically translate to having the poor secure in their access to food. Also, world food level prices do not necessarily reflect local pricing, especially for developing countries (Sande 2008).

Food is a dilemma particularly for people earning less than USD 2 a day. It can be verified that the big problem concerning bioenergy is the food dilemma, but specifically of people that have an income of up to USD 2 a day. In this context, the notion of a "right to food" can arise (Oenema 2008).

Monbiot (2005) and Rosillo-Calle and Johnson (2010) affirm that despite the existence of 80 million people who are permanently malnutritioned, the global increase of grain production is being used to feed animals. The number of animals in the planet has quintupled since 1950. And the main reason is that those who buy products originated from milk and meat cattle have larger bargaining power than those that only purchase food for subsistence. Monbiot (2005) considers green oil both a disaster to humanity as well as for the environment.

In his studies, Mol (2007; 2008) affirms that less than 3.5 percent of world food production in 2007 was used for bioenergy production. Since the EU required that transportation oil should contain at least 5.75 percent of biofuels until the end of 2010, there have been signs of competition between grains and energy. In 2007, 4.5 million tons of grains were processed with bioethanol. But the grain supply in the entire world has decreased substantially. Mexico was one of the countries that suffered with the increase of the price of corn due to the fact that this product is imported from the United States, which jeopardized the production of tortillas (Furfari 2008).

In 2007, grain prices in the EU were of 160 to 260 euros per ton, which resulted in a smaller harvest, thus generating speculations (Mol 2008). According to Faaij (2010), the sustainable offer of biomass is vital for any bioenergy production market activity. Given the high expectations for bioenergy in a global scale, many nations have been pressuring for the availability of biomass resources. Due to the high prices of fossil oil, the competitiveness of the use of biomass has greatly increased.

Dijk (2008) pointed out that only 1 percent of the arable lands of the world is used for the production of bioenergy, and this fact illustrates that the recent prices of agricultural commodities cannot be a result of the substitution of biomass for food production to biomass for oil production. The prices of many commodities, including energy, metals, and minerals, as well as agricultural commodities have risen above 5 percent per year as a result of population growth and continuous global economic growth.

The United States, China, and Brazil, as well as the EU can direct the food production and land use with some governmental instruments, as it is easier to guide national markets than the international market. But these actions have made food and bioenergy global commodities for entrepreneurs and investors. However, another problem was created concerning subsidies especially in the

United States and Europe. The United States subsidizes its biofuel producers in order to reduce the dependence of fossil oil importation, while Europe subsidizes their producers as an answer to climate problems (Mol 2008).

Sande (2008) commented on Europe, which has already decided to make obligatory the inclusion of biofuels in transportation oils to replace fossil fuels. This decision was made to reduce the emission of greenhouse effect gases and attempt to rectify climate changes. The goal is to include 5.75 percent of biofuels in 2010 and 10 percent in 2020, provided that this percentage can be reached through available sustainable paths without negatively affecting the opportunities of small producers.

According to Hass, Larivé, and Mahieu (2003), the emission of greenhouse effect gases in the atmosphere by biofuels, biodiesel, and ethanol in the EU represents a reduction of between 15 and 70 percent from diesel oil. In Brazil, the production of alcohol is associated with more than 90 percent reduction (Alckmin and Goldemberg 2004).

The EU Council created the biofuel law. Under the Council's mandate, the European Commission sent the proposal to the Parliament and the Council. It is anticipated that it will be enacted by the end of 2010 (*Gazeta Mercantil* 2008).

The production of bioenergy is a delicate matter for discussion because it affects the production of biomass for food. Consumers of fresh foods have questioned the origin, impact on the food supply, and sustainability dimensions of these provisions. As with foods, new seeds cannot be introduced without meeting sustainability criteria (Dijk 2008).

The *World Development Report* (World Bank 2007) concerning agriculture reveals that bioenergy is a potential renewable energy resource and provides possible new markets for agricultural producers. However, few bioenergy projects are economically viable and most have social and environmental costs, such as pressures on food prices, intensification of the competition for land and water, and on occasion, deforestation (Sande 2008).

Oenema (2008) mentions limitations of bioenergy in Brazil pointed out by the Dutch Inter-Church Organization for Development Co-operations (ICCO):

1. Environmental risks due to large-scale agricultural monoculture;
2. Social risks due to unsatisfactory work conditions in the fields; low salaries and land reform;
3. Competition with food grains (this discussion presents some factors that should be considered, such as the decrease of world food supply due to environmental problems and the increasing demand for food in Asian countries).

However, Dijk (2008) uses three arguments for investing in the production of sustainable biomass for food, seeds, and oils: the need to integrate sustainable practices in future global agricultural production, the need to accelerate agricultural productivity, and the many benefits for production of biomass for rural development.

Read (2005) affirms that the use of bioenergy can reduce the presence of CO_2 in the atmosphere from about 375ppm and to around 280ppm in 2060. This can also contribute to the realization of the Kyoto Protocol.

Moreira (2005) asserts that the demand for more food can result in more job opportunities in rural areas to grow it, and that it is necessary to understand biomass production should be profitably undertaken by the poor rather than merely used by the poor. Through this immense new opportunity for producers, the associated income increase could reduce the number of poor people in the world by at least 200–300 million.

Similarly, Read (2005) postulates that grain production for biomass can replace half the oil consumption in the world, generate half the electricity demand, create 300 million new jobs, and significantly reduce the emission of greenhouse gases.

In the twentieth century, there will probably not be only one overarching source of energy, as during the nineteenth century with coal and in the twentieth century with oil, although there is still much to be accomplished to get to that point. Several energy sources will coexist; renewable sources should become predominant with a smaller proportion of polluting sources. Those with biological origins particularly should greatly expand in the coming decades (Smith and Taylor 2008).

The Growing Prominence of Stakeholders

Influence generated by the interests of various stakeholders distributed around the world can initiate actions in organizational, political, legislative, and civil areas. However, various stakeholders present differing interests, values, and cultures, which might suggest that viewing a dialogue with stakeholders of bioenergy might not be adequate for addressing climate change. On the other hand, it could be an opportunity for the new comprehension mechanisms to connect the goals of the stakeholders in common actions to actualize the vision of global sustainability as a business imperative.

The Promise of Dialogue

Dialogue can be a key tool for solving problems. Unfortunately, there has been too little dialogue with the stakeholders of bioenergy. Examples of using dialogue to find solutions are put forth by Sande (2008). Though biofuel production is clearly viable and has stark economic benefits, people and nations have to discuss its attendant issues to find common ground and clarify where they have divergent goals et cetera.

The problems faced by stakeholders and organizations that seek environmental sustainability include noncontinuity of implemented processes, lack of participation of certain stakeholders, a lack of valuation of environmental management processes, and lack of connections between stakeholders. These factors constrain the use of dialogue and rather promote conflict to the detriment of environmental preservation.

An example is the dialogue between stakeholders concerning bioenergy and food supply issues that was promoted in The Netherlands during the On World Food Day event on October 16, 2007, on the focal topic of biomass for food and oil. The Dutch also have a committee created by their government to make recommendations concerning the sustainability of the use of biomass for energy purposes.

Another example occurred in 2007 in the city of Bangkok, Thailand, at a meeting organized by Organization of the Petroleum Exporting Countries (OPEC) called the Multilateral Dialogue for the Development of Future Energy Workshop (*OPEC Bulletins* 2007). This session sought to foster listening and action supporting Asian energy demand.

In the United States, a Harvard University workshop was held on May 9, 2007, on future implications of bioenergy for economic development and the international market with a dialogue among academics, international institutions, governments, and the private sector to explore the implications of the emergence of the bioenergy market, mainly after the target announced by President George W. Bush to increase the use of biofuel by 5 percent by 2017 (Lee et al. 2007).

These examples demonstrate that the use of dialogue could be an alternative to solve the existing conflicts about bioenergy related to food dilemmas. How can dialogue among various stakeholders find solutions for climate change through the use of bioenergy? The opportunity to use bioenergy and the dilemma concerning food production has become a great opportunity for the use of dialogue between every stakeholder in search of alternatives for this conflict.

The Dialogues Theory: Perspective for Global Sustainability

The participation of the stakeholders in environmental debates has favored the discussion of new themes locally, nationally, and globally (NAE 2005). Thus, concern for the environment has progressively achieved more legitimacy among countries. It is important to point out that individual behavior represents a critical factor of global climate change. Therefore, blame should not only be given to industrial pollution, government failure, or institutional inefficiency. (Figueres and Ivanova 2005).

In Stakeholder Theory (Freeman 1984), stakeholders are framed as individuals or organizations that affect or are affected by an organization's objectives or problems. The stakeholders of a business firm have a much wider scope for acting than the firm itself does, say for addressing climate factors (Key 1999). Stakeholder Theory is grounded in attaining justice, equity, and social interchange among those who have an interest in a firm. Stakeholders need to be identified, their powers and influence mapped, and their input used to possibly refine the firm's initial objectives and strategies. Both political and scholarly approaches to business approaches to managing issues related to how it impacts the natural environment shift when dialogues with stakeholders ensue on this topic (Kloprogee and van der Sluijs 2006). Dialogue based on scientific knowledge of the natural environment can bring researchers and organization's stakeholders closer (Welp et al. 2006a; Welp and Stoll-Kleeman 2006). Business managers can find that stakeholders have know-how that can importantly support their endeavor to comprehend, represent, and analyze global environmental changes.

There are four compelling reasons for a business firm to enter into a dialogue with its stakeholders: (1) its stakeholders can play an important role in identifying relevant social issues and can usefully influence agendas, (2) feedback from stakeholders assists researchers in their ongoing improvement of their research methods and models and, thus, the quality of the research results, (3) stakeholders' perspectives on germane ethical issues can suggest ways the company might include the impact of changing global conditions in its research, and (4) stakeholders can alert businesses to important data and knowledge to which they might not have been privy (Welp et al. 2006).

This approach allows dialogues as interfaces between different areas of expertise on climate change and agribusiness. The main types of guided dialogues are: political dialogues (Innes and Booher 2003), multistakeholder dialogues for governments (Hemmati 2007), and corporate dialogues (Jesper 1998). A nonhierarchical structure in a relationship with multiple stakeholders can overcome issues associated with different rates of collaboration by different stakeholders (Streck 2005).

Agribusiness Stakeholders in Climate Change and Bioenergy for Global Sustainability

In order to introduce a framework, it is important to review some dialogue issues. Dialogues should include the diversity of stakeholders and should be framed to enable exchanges between technical and ordinary citizens (Rorty 1991).

Transdisciplinary research can solve complex problems such as biodiversity loss and climate changes mainly where particular discipline research independently fails to succeed (Pohl 2005). This line of thought introduces the relevance of stakeholders in agribusiness, since the collaboration between businesses, industries, research institutions, and communities is likely to develop better approaches to address environmental issues.

The insertion of stakeholders into an agribusiness environment with a focus on addressing climate change and bioenergy bolsters the chance for realism.

Stakeholders can be main agents and guides of change processes of agribusiness sectors. The dialogues between agribusiness stakeholders concerning climate changes and bioenergy are not static, and this is the reason the three variable sets (stakeholders, agribusiness, and climate changes) should be systematically studied.

Convention Theory suggests approaches to coordination where all of the important factors are not clear to many of the actors. It deals with collective rules that interpose themselves when various stakeholders try to work out something in a situation that becomes clearer through dialogue. Among the types of conventions (rules or models) that might be used are: *civic conventions* where stakeholders mutually perceive a common interest or an overriding objective that transcends their individual interests; *market conventions* where stakeholders respond to the demands of external markets; and *domestic conventions* presenting relations of trust and loyalty among stakeholders and the need to help each on the basis of their personal knowledge of each other (Chevassus-Loza and Valceschini 1994).

Brazil has used a market convention in assessing impacts on the natural environment. Concerns have centered on maintaining its international markets. There are economic rationales for some stakeholders related to the search for raw materials (natural resources) by those organizations, which are essential to their products and survival. Therefore, they look to conventions (models of behavior by similar firms, for example) that justify their use of natural resources. Networks of stakeholders might suggest ways to mitigate ecological harm within the market context that a firm finds itself in. Stakeholders enter dialogues with each other with interests and perspectives that vary according to their individual interests and the region they are located in.

Final Conclusion

The search for renewable energy, in this case bioenergy, has become a strategy for the development of organizations, society, and stakeholders. The huge current use of nonrenewable sources in the world energy supply gives society the challenge of focusing on the search for alternative sources of energy. In this sense, the comprehension of the use of the dialogues between stakeholders might lead to better alternative solutions to conflicts to the benefit of the environment and organizations. The objective of this chapter has been to analyze conflicts involving bioenergy and food and how the dialogues have played out to find out a common understanding among stakeholders.

We have shown different conflicts and conventions created for stakeholders of Brazil. The risk is about arable land and intense competition over the production of bioenergy and food. Brazil has a unique approach to sustainable development of bioenergy.

References

Agroanalysis. 2007. Abastecimento: Estoques mundiais de alimentos em baixa. *Agroanalysis* 27 (12): 29–30.

Alckmin, G., and J. Goldemberg. 2004. Assessment of greenhouse gas emissions in the production and use of fuel ethanol in Brazil. http://www.unica.com.br (accessed July 20, 2008).

Cerri, C. 2009. Brazilian greenhouse gas emissions: The importance of agriculture and livestock *Scientia agricola* 66(6): 831–843.

Chevassus-Loza, E., and E. Valceschini. 1994. Les concepts de l'economie de conventions et leur articulation [The concepts of the economics of conventions and their articulation]. Paper presented at CIRAD Conference on Institutional Economics. Montpellier, France. Summary notes. http://www.msu.edu/course/aec/932/Convention_theory_notes.pdf (accessed April 30, 2010).

Dijk, D. 2008. Why not the right for food and fuel? In *Bio-fuels and food security: Dialogue among stakeholders on dilemmas about biomass for food and/or fuel,* ed. P. S. Bindraben and R. Pistorius, 17–20. Report, 167. Wageningen, The Netherlands: Plant Research International.

Faaij, A. 2010. A roadmap for biofuels in Europe. *Biomass and Bioenergy* 34(2): 157–158.

FAO. 2008. Food. http://www.FAO.org.br/faq_alimentos.asp (accessed July 15, 2008; no longer accessible).

———. 2008a. *Low-income food-deficit countries.* http://www.FAO.org/focus/e/speclpr/lifdcs.htm (accessed July 20, 2008).

Figueres, C., and M. H. Ivanova. 2005. Mudanças climáticas: interesses nacionais ou um regime global? In *Governança ambiental global: opções e oportunidades* [*Global environmental governance: Options and opportunities*], ed. D. Esty and M. H. Inavona, 233–255. São Paulo: Senac.

Freeman, R. E. 1984. *Strategic management: A stakeholder approach.* Boston: Pitman/Ballinger.

Furfari, A. 2008. *Biofuels: Illusion or reality? The European experience*. Paris: Technip.

Gazeta Mercantil. 2008. Uso de biocombustível na UE é ainda motivo de controvérsia: A lei de biocombustível se deve a uma decisão do Conselho da UE relacionada à mudança climática e não energética. (July 3). http://www.ecodebate.com.br/2008/07/03/uso-de-biocombustivel-na-ue-e-ainda-motivo-de-controversia/ (accessed May 1, 2010).

Griffin, J. 2007. Biofuel: Food for thought. *OPEC Bulletin* 38(5–6): 34–39.

Hass, H., J.-F. Larivé, and V. Mahieu. 2003. *Well-to-Wheels Analysis of future automotive fuels and powertrains in the European context*. http://www.ies.jrc.cec.eu.int (accessed August 13, 2008; currently not accessible).

Hemmati, M. 2007. *Participatory dialogue: Towards a stable, safe and just society for all*. Report commissioned by UN DESA Department for Social Policy and Development. New York: United Nations.

Innes, J. E., and D. E. Booher. 2003. Collaborative policy making: Governance through dialogue. In *Deliberative policy analysis: Governance in the network society*, ed. M. W. Hajer and H. Wagenaar, pp. 33–59. Cambridge, UK: Cambridge University Press.

Jesper, G. 1998. Corporate legitimacy in risk society: The case of the Brent Spar. *Business Strategy and the Environment* 7: 213–222.

Key, S. 1999. Toward a new theory of the firm: A critique of stakeholder "theory." *Management Decision* 37(4): 317–328.

Kloprogee, P., and J. P. van der Sluijs. 2006. The inclusion of stakeholders' knowledge and perspectives in integrated assessment of climate change. *Climatic Change* 75(3): 359–389.

Lee, H., W. Clark, R. Lawrence, and G. Visconti. 2007. Implications of a future global biofuel market for economic development and international trade. In *Workshop on the future implications of a global biofuels market for economic development, environment and trade*. May 9. Cambridge, MA: John F. Kennedy School of Government, Harvard University. http://www.hks.harvard.edu/var/ezp_site/storage/fckeditor/file/pdfs/centers-programs/centers/cid/ssp/docs/events/workshops/2007/biofuel/Harvard_Biofuels_Workshop_report_070509.pdf (accessed May 1, 2010).

Metz, B. 2010. *Controlling climate change*. Cambridge, UK: Cambridge University Press.

Mol, A. P. J. 2007. Boundless biofuels? Between environmental sustainability and vulnerability. *European Society for Rural Sociology* 47(4): 297–315.

———. 2008. *Bioenergy: A growing market in need of direction. Wageningen Update* 1: 8–11.

Monbiot, G. 2005. Worse than fossil fuel. *Guardian* (December 6). http://www.monbiot.com/archives/2005/12/06/worse-than-fossil-fuel/ (accessed April 28, 2010).

Moreira, J. R. 2005. Agreeing and disagreeing. *Renewable Energy for Development* 18(2): 7.

NAE (Núcleo de Assuntos Estratégicos da Presidência da República). 2005. *Mudança do clima.* Vol. 1. Brasília: NAE, February 2005. (Cadernos NAE, n. 3). http://www.sae.gov.br/site/?p=2781 (accessed May 1, 2010).

Oenema, S. 2008. Bio-fuel viewed from a perspective of the right to food. In *Bio-fuels and food security: Dialogue among stakeholders on dilemmas about biomass for food and/or fuel,* ed. P. S. Bindraban and R. Pistorius, 37–39. Plant Research International B. V., Wageningen, Report 167.

OPEC Bulletins. 2007. Bangkok energy dialogue workshop looks at strategies for dealing with challenging years ahead: Multilateral dialogue seen as essential for developing sound and orderly energy future, 5–7: 12–19.

Pohl, C. 2005. Transdisciplinary collaboration in environmental research. *Futures* 37: 1159–1178.

Porter, M. E., and C. van der Linde. 1995. Green competitive: Ending the stalemate. *Harvard Business Review* 73(5): 120–134.

Read, P. 2005. Food for thought—world trade in biofuel offers sustainable food supply and much more. *Renewable Energy for Development* 18(2): 4–5.

Rorty, R. 1991. Solidarity or objectivity? In *Objectivity, Relativism and Truth,* ed. R. Rorty, 21–34. Cambridge, UK: Cambridge University Press.

Rosillo-Calle, F., and F. X. Johnson, eds. 2010. *Food versus fuel: An informed introduction to biofuels.* London: Zed Books.

Sande, T. V. 2008. Opportunities and threats bio-fuel for agriculture and rural development. In *Bio-fuels and food security: Dialogue among stakeholders on dilemmas about biomass for food and/or fuel,* ed. P. S. Bindraban and R. Pistorius, 25–28. Plant Research International B. V, Wageningen, Report 167.

Smith, Z. A., and K. D. Taylor. 2008. *Renewable and alternative energy resources: A reference handbook.* Santa Barbara, CA: ABC-CLIO.

Streck, C. 2005. Redes Globais de políticas publicas como coalizões para mudança. In *Governança ambiental global: opções e oportunidades,* ed. D. Esty and M. Inavona, 139–159. São Paulo: Senac.

Welp, M., A. C. de la Vega-Leinert, S. Stoll-Kleemann, and C. Fürstenau. 2006. Science-based stakeholder dialogues in climate change research. In *Stakeholders dialogues in natural resources management,* ed. S. Stoll-Kleemann and M. Welp, 213–240. Heidelberg: Springer-Verlag.

Welp, M., and S. Stoll-Kleemann. 2006. Integrative theory of reflexive dialogues. In *Stakeholder dialogues in natural resources management,* ed. S. Stoll-Kleemann and M. Welp, 43–78. Heidelberg: Springer-Verlag.

World Bank. 2007. *World development report. Agriculture and development. 2008.* Washington, D.C.: World Bank.

Yeganiantz, L., E. Contini, E. Cruz, R. Brandini. 1984. *An antipollution stance of Germany and potential market for Brazilian alcohol in the European Community.* Brasília: Embrapa.

Yeganiantz, L., J. R. Alencar, J. M. Nogueira, E. Contini. 2007. Sustainable biofuel in Brazil and the United States of América. Proceedings of VI International PENSA Conference, Ribeirão Preto. Brazil.

Index